THE OIL WARS MYTH

THE OIL WARS MYTH

Petroleum and the Causes of International Conflict

Emily Meierding

CORNELL UNIVERSITY PRESS ITHACA AND LONDON

First published 2020 by Cornell University Press

Printed in the United States of America

Library of Congress Cataloging-in-Publication Data

Names: Meierding, Emily, author.
Title: The oil wars myth : petroleum and the causes of international conflict / Emily Meierding.
Description: Ithaca [New York]: Cornell University Press, 2020. | Includes bibliographical references and index.
Identifiers: LCCN 2019035806 (print) | LCCN 2019035807 (ebook) | ISBN 9781501748288 (hardcover) | ISBN 9781501748950 (pdf) | ISBN 9781501748943 (ebook)
Subjects: LCSH: Petroleum industry and trade—Political aspects—History—20th century. | World politics—20th century. | War—Causes. | Politics and war.
Classification: LCC HD9560.6 .M44 2020 (print) | LCC HD9560.6 (ebook) | DDC 338.2/7280904—dc23
LC record available at https://lccn.loc.gov/2019035806
LC ebook record available at https://lccn.loc.gov/2019035807

The most stubborn facts are those of the spirit, not those of the physical world.

—Jean Gottmann

Contents

Tables and Maps

Tables

Maps

Acknowledgments

This book began, I suppose, with a duck. Scrooge McDuck, to be exact, whom I discovered one Christmas at my grandmother's house in Ventura, California. Her trove of comics, written by the "Duck Man," Carl Barks, introduced me to the stories of El Dorado, the Seven Cities of Cibola, and the Golden Fleece. I devoured them while sitting on the floor of a living room whose windows overlooked the distant lights of oil rigs gleaming in the Santa Barbara Channel.

On the pathway from that preliminary encounter to this book's completion, I have accumulated innumerable personal and professional debts, for which I can only begin to express my gratitude. I will also inevitably look back at these acknowledgments and be horrified by at least one omission. Whoever you are, please track me down and insist that I buy you a drink.

Thanks go to my advisers at the University of Chicago—Charles Lipson, John Mearsheimer, and Steven Wilkinson. Thanks also to my colleagues at the Graduate Institute of International and Development Studies, especially Liliana Andonova, Susanna Campbell, Stephanie Hofmann, and Annabelle Littoz-Monnet, as well as to Kathryn Chelminski, Daniel Norfolk, Joanne Richards, and Alain Schaub for their research support. Thanks to my new colleagues at the Naval Postgraduate School, including Jason Altwies, Naaz Barma, Erik Dahl, Chris Darnton, Diego Esparza, Covell Meyskens, Afshon Ostovar, Jessica Piombo, and Rachel Sigman, who have made working in Monterey even more of a pleasure. Additional thanks to my students in Energy Security and Geopolitics (fall 2018), who stepped in where my pop culture knowledge failed.

I also thank all the energy experts whose engagement, encouragement, and criticism have made this a much better book (and who bear no responsibility for its shortcomings): Jeff Colgan, Eugene Gholz, Charlie Glaser, Kathleen Hancock, Rose Kelanic, Philippe Le Billon, Jonathan Markowitz, Victor McFarland, Mark Nance, Shannon O'Lear, Ben Smith, Roger Stern, Adam Stulberg, Thijs Van de Graaf, and Bob Vitalis. I want to express particular thanks to Michael Klare for his extraordinary graciousness; when this book is criticized, you will be my model of how to respond.

Thanks to the researchers associated with the now sadly shuttered Conflict Records Research Center at the National Defense University, especially David Palkki, for giving scholars access to an invaluable resource. Further gratitude goes to the librarians at the Naval Postgraduate School, the Graduate Institute of

International and Development Studies, Stanford University, the University of California, Berkeley, and the University of Chicago, especially those who had to deal with my Interlibrary Loan and microfiche requests. Thanks to my production editor, Kristen Bettcher, and the team at Cornell University Press, especially Roger Haydon, for supporting a book that raced from rationalism to critical geopolitics with possibly reckless abandon. Thanks also to Mike Markowitz for the book's fantastic maps.

Enormous gratitude goes to all the other friends who have supported me, practically and emotionally, along the way, including Jon DiCicco, Dan Drezner, Andrea Everett, Anne Holthöfer, Jennifer London, Patchen Markell, Alex Montgomery, Owen O'Leary, Nathan Paxton, Negeen Pegahi, Keven Ruby, John Schuessler, Scott Siegel, and Nick Smith. Particular shout-outs go to those who played key roles in the home stretch: Jenna Jordan, for an inspirational conversation at Whole Foods; Frank Smith, who unknowingly helped me write my introduction; Jon Caverley, who encouraged me to celebrate early and often; and Jill Hazelton and Karthika Sasikumar, for their liberality with the "like" button.

Special thanks, of course, to Harper House's Chris Buck, Andrew Dilts, and Sina Kramer, who gave me my first tastes of scotch, key lime pie, and *The West Wing*. To my parents, Mark and Linda, and my sister, Julie, for their love, support, and inspiration. To Louise for always, always believing in me. And to Louis, who has played the roles of geological consultant, cheerleader, copyeditor, co-expat, and partner with equal generosity and élan. Je t'aime, mon amour.

THE OIL WARS MYTH

Introduction

BLOOD AND OIL

Oil powers modern life. It provides gasoline for our cars, diesel for our trucks and trains, heavy oil for our ships, and jet fuel for our planes. Without oil, transportation would come to a standstill. Recreational and business travel would collapse. Shipping would cease. Military vehicles would be incapacitated; from jeeps to destroyers, modern defense establishments run on petroleum-based fuels.[1] Even vehicles that consume biofuels, compressed natural gas, or nuclear energy would literally grind to a halt; without lubricants, derived from oil, there would be nothing to grease their gears.

Absent oil, New Englanders would experience very cold winters, as many communities use petroleum as a heating fuel. In Hawaii, the one US state that still relies on oil to produce electricity, losing it would turn out the lights. Outside the United States, oil-fueled power plants would shut down and people would no longer be able to compensate by firing up diesel-fueled generators. Globally, the petrochemical industry would take a beating. Plastics, fertilizers, paints, synthetic fabrics, and medicines are all manufactured from hydrocarbons. Absent petroleum feedstocks, output would drop, dramatically.

Recognizing oil's exceptional value, Henry Kissinger called it "the world's most strategic commodity," while the international relations theorist Hans Morgenthau labeled oil "the lifeblood of industrially advanced nations."[2] Petroleum resources are also vital to the countries that produce them. If their oil disappeared, states like Saudi Arabia, Russia, and Venezuela would lose a crucial revenue stream. They would no longer be able to balance their budgets and maintain social spending, inviting popular discontent. The French industrialist Henri Bérenger summarized petroleum's extraordinary contributions to national power and prosperity in 1919 by proclaiming that "he who owns the oil will own the world, for he will rule the sea by means of the heavy oils, the air by means of the ultra-refined oils, and the land by means of the petrol and the illuminating oils. And in addition to these he will rule his fellow men in an economic sense, by reason of the fantastic wealth he will derive from oil."[3]

Oil's unique value has led many people to assume that countries fight to obtain it. "Oil wars" figure prominently in popular culture. The final season of the critically acclaimed television show *The West Wing* depicted Russia and China poised on the brink of war over oil fields in Kazakhstan. As one character stated, "The Russians want the oil and are willing to fight for it. . . . Worst-case scenario, two nuclear powers are positioning themselves to actively engage in an armed conflict over oil."[4] In the speculative novel *Ghost Fleet*, a US naval officer explains that "the scramble for new energy resources . . . [is] sparking a series of border clashes around the world." A few chapters later, China initiates World War III in order to grab a natural gas field in the US Exclusive Economic Zone.[5]

These fictional storylines echo academic assertions that countries fight for petroleum resources. Andrew Price-Smith claims that "oil reserves are often a target of conquest," while Christopher Fettweis contends that "resources have always been a great motivator for war."[6] Michael Klare predicts future resource wars, averring, "That conflicts over oil will erupt in the years ahead is almost a foregone conclusion."[7] Journalists also peddle oil war narratives; whenever international competition heats up in petroleum-rich regions like the Arctic, the East China Sea, the South China Sea, or the eastern Mediterranean, reporters predict that it will escalate to outright, militarized violence. Politicians have also reinforced the belief that countries fight for oil. As early as World War I, French prime minister Georges Clémenceau purportedly stated that "a drop of oil is worth a drop of blood."[8] In addition, generations of antiwar protesters have implied a connection between petroleum and conflict with their persistent invective, "No blood for oil."

But do countries actually fight wars to acquire oil resources? We assume that we know the answer to this question: yes. However, this book will demonstrate that the conventional wisdom is, at best, a dramatic exaggeration and, at worst, a complete fiction.

Oil Wars: Underdefined and Underresearched

To determine whether oil causes international conflict, we first need to clarify the meaning of *oil wars*. The term is often deployed broadly to refer to any type of militarized conflict that is linked—in any way—to oil. In *Resource Wars*, Michael Klare spans different scales of conflict, examining oil-related civil wars along with oil-related interstate violence.[9] Jeff Colgan identifies eight distinct causal pathways from petroleum to international conflict, ranging from conquest of oil resources, to fights over petroleum transportation routes, to externalization of oil-oriented civil wars, to conflicts aimed at preventing the consolidation of control over global petroleum supplies.[10] André Månsson also distinguishes between conflicts to secure global energy flows and conflicts over natural resource ownership.[11]

Some forms of oil wars have attracted significantly more academic attention than others. Research on oil and civil wars, for example, is expansive. Since scholars began to focus on the topic in the mid-1990s, hundreds of books and articles have attempted to ascertain whether and how oil resources contribute to intrastate instability and violence.[12] Some researchers argue that oil creates immediate incentives and opportunities for contention, while others emphasize the "slow violence" caused by petroleum extraction.[13] The phenomenon of "petro-aggression" has also generated a substantial body of research. In this pathway from oil to conflict, petroleum-producing states with revolutionary governments use their resource revenue to facilitate international aggression, usually against their neighbors, who may or may not possess oil resources.[14]

In contrast, other linkages between oil and violence have attracted limited systematic scholarly attention. Surprisingly, one of these understudied pathways is the type of oil-related contention that most often captures the popular imagination: what I call the "classic oil war."[15] These are severe militarized interstate conflicts in which participants fight to obtain petroleum resources. One country may attempt to grab another's oil. Or countries may forcefully contest authority over disputed resource reservoirs. Other researchers have alluded to this form of contention in their overviews of oil-related violence. Colgan labels this the "resource war" pathway, Charles Glaser calls these conflicts "classic resource wars," and Georg Strüver and Tim Wegenast identify them as "classical resource wars."[16] Indra de Soysa, Erik Gartzke, and Tove Grete Lie assess the related "blood oil" hypothesis, which proposes that "petroleum is a prize over which nations will fight."[17] I use the term *classic oil wars* to synthesize these concepts.

Classic oil wars are widely assumed to be regular and dangerous events in the modern international system. Historians and political scientists have classified a number of major conflicts, including Japanese and German aggression in World War II (1939–1945), Iraq's invasion of Kuwait (1990), the Iran–Iraq War (1980–1988), the Falklands War (1982), and the Chaco War (1932–1935), fought by Bolivia and Paraguay, as classic oil wars. Some less prominent international conflicts, including Cameroon and Nigeria's Bakassi Peninsula confrontations and

Ecuador and Peru's contention over the Maynas/Oriente region, have also been identified as oil-driven.[18]

However, scholars also describe classic oil wars as an "underdeveloped research agenda" and acknowledge that "systematic empirical evidence about the frequency of such conflicts is lacking."[19] Only a handful of statistical analyses have evaluated oil resources' connection to interstate conflict, and those produced inconsistent results, with some finding that petroleum endowments increase the likelihood of contention, while others did not.[20] Our understanding of classic oil wars is therefore strikingly limited, especially for events that are widely assumed to be a prominent and threatening feature of international politics.

Classic Oil Wars: Fact and Fiction

This book argues that countries' willingness to fight for oil resources has been overstated. Oil is an exceptionally valuable commodity. All countries need petroleum to power their militaries and economies, while oil-producing states also depend on resource revenue to balance their budgets. Consequently, all countries would prefer to control more oil and will suffer severely if their petroleum needs are not met. Still, classic oil war believers make an unmerited cognitive leap by jumping from the observation that oil is extremely valuable to the conclusion that countries fight to acquire it.

I argue, to the contrary, that countries avoid classic oil wars. Despite petroleum's extraordinary utility, international resource grabs do not pay. Seizing and exploiting foreign oil is far more challenging than oil war believers have recognized. At a minimum, states face four sets of impediments to obtaining the petroleum resources and revenue they desire. First, an aggressor must invade foreign or contested territory to seize oil fields. In the process, it may damage the oil reservoirs and infrastructure it aims to acquire. Second, to exploit captured resources, the aggressor must occupy conquered territories over the long term, exposing its forces and the appropriated petroleum industry to a hostile local population, which will try to impede resource exploitation.

Third, aggressors face retaliation from the international community, which regards oil grabs as illegitimate acts, as well as threats to international security. Third-party states and international organizations will therefore oppose classic oil wars diplomatically, economically, and militarily by censuring aggressors, sanctioning their oil sales, or forcefully ejecting them from conquered territories. Fourth, aggressors will have difficulty attracting the foreign investment they may require to exploit captured petroleum resources, as oil companies are reluctant to participate in petroleum projects in areas where political authority is illegitimate or uncertain. Collectively, these four sets of impediments—which I call invasion, occupation, international, and investment obstacles—dramatically diminish the petroleum payoffs of classic oil wars. Recognizing these limitations, states refrain from launching major conflicts to grab petroleum resources.

Yet, if countries avoid classic oil wars, why do we continue to believe in these conflicts? Why do we predict that international competitions over petroleum resources will escalate into major confrontations? Why do we assume that countries' oil ambitions are a significant cause of international contention? Why do we uncritically accept classic oil war storylines in popular novels and television shows? And why has this topic attracted so little rigorous scholarly attention?

The answer to these questions, I argue, is that classic oil wars are a myth: a foundational story we tell ourselves about how the world works. I invoke the terminology of myths to describe our collective belief that countries fight wars to acquire petroleum resources, because it highlights these conflicts' unusual epistemological status. Classic oil wars are not a typical social scientific concept: clearly defined, carefully theorized, and regularly subjected to empirical assessments. Instead, they are taken for granted; they are simply presumed to be a prominent feature of international politics. Put another way, we take classic oil wars' existence on faith. As a result, these conflicts have been immune to theoretical critiques and empirical challenges. The oil wars myth creates a collective intellectual blind spot.

This misplaced conviction is problematic, for academic and policy reasons. Oil war beliefs shape contemporary foreign policy choices. The United States established its military presence in the Persian Gulf in order to deter international petroleum grabs.[21] Polar states are expanding their forces and facilities in the Arctic, partly to defend their oil resources against foreign aggression. China's activities in the South and East China Seas, including confrontations with its neighbors over oil and gas exploration, as occurred near the Paracel Islands in 2014, look more threatening because we assume that competitions involving hydrocarbon resources can spiral into severe interstate conflicts. However, if states actually avoid classic oil wars, these confrontations are not a serious security threat. Nor do states need to be as vigilant in defending their own, or other states', petroleum endowments. In addition, even if oil prices rebound from their current slump, we do not need to fear great power petroleum wars. Discovering that states do not engage in international oil grabs could therefore fundamentally alter our foreign policy choices.

Academically, determining that states avoid international petroleum grabs would change our answer to a core international relations question: What causes war? Bruce Russett has summarized the conventional wisdom about resources and interstate violence by asserting that "the need for assured access to raw materials . . . provides a powerful driving force towards present-day international conflict."[22] Of all natural resources, oil is presumed to be the most likely to incite contention, as a result of its exceptional economic and military utility. If this proposition is false, however—if countries' oil aspirations are only a marginal motive for international conflict or provoke only minor confrontations—then we need to update our understandings of resources' contributions to interstate war, territorial disputes, and other forms of conflict. We should also ask ourselves, if countries have not been fighting for oil resources, what have they been fighting for? What has the oil wars myth obscured?

Method and Plan of the Book

This book dissects classic oil wars using a three-step approach. First, I examine the intellectual history of the oil wars myth to explain why we are so eager to believe that states fight to control petroleum resources. Second, I interrogate classic oil wars theoretically, showing that the causal logic underpinning these claims does not hold up to critical scrutiny. Third, I assess these conflicts empirically to determine whether, in the words of energy scholar David Victor, "classic resource wars are good material for Hollywood screenwriters. They rarely occur in the real world."[23] My analysis closely examines many of the historical conflicts that are commonly identified as classic oil wars—World War II, the Iran–Iraq War, the Chaco War, the Falklands War, and Iraq's invasion of Kuwait—and presents a broader assessment of over six hundred militarized interstate disputes (MIDs), from 1912 to 2010, to determine whether oil ambitions contributed to leaders' decisions for international aggression.

Since belief in classic oil wars rests on petroleum's value, chapter 1 begins by tracing oil's development from a largely invisible, worthless material into a vital natural resource. It then discusses oil's contemporary military and economic significance for petroleum-exporting and petroleum-importing states. The chapter also situates the book within an existing international relations debate about the value of conquest, showing how oil's exceptional utility has led most theorists to assume that the resource is worth fighting for. As I note, oil war skeptics have questioned these claims. However, their piecemeal critiques and limited empirical challenges have failed to dislodge classic oil war beliefs. To explain the tenacity of these convictions, the chapter introduces the idea of the oil wars myth: a powerful story we tell ourselves about countries' willingness to fight over petroleum resources.

Chapter 2 explains why we believe in classic oil wars. It argues that the credibility of the oil wars myth arises from its connection to two other hegemonic narratives about the causes of violent conflict: Mad Max and El Dorado. The Mad Max myth asserts that countries, groups, and individuals fight because of existential need; if they fail to acquire more vital raw materials, they will die. The El Dorado myth intimates that actors fight out of greed; they grab resources in order to gain extraordinary wealth. The chapter traces these dual narratives across centuries of academic and popular discourse. It also shows how the Mad Max and El Dorado myths were applied to oil after it was identified as a valuable natural resource. Because classic oil wars align with these two well-established narratives, they appear eminently plausible. The oil wars myth is easy to believe in.

Yet the myth rests on shaky logical foundations. Chapter 3 challenges a key assumption underpinning classic oil war claims. It argues that, although oil war believers correctly observe that petroleum is extremely valuable, they err in jumping to the conclusion that fighting for oil pays. I describe four sets of impediments to seizing and exploiting foreign oil resources, all of which reduce the petroleum payoffs of classic oil wars: invasion, occupation, international, and investment obstacles. The intensity of each set of obstacles varies temporally and geographically. However, even under favorable conditions, the petroleum payoffs of international aggression are far lower than classic oil war believers have assumed. As a result, states are likely to avoid fighting for petroleum resources.

Chapter 4 provides a preliminary evaluation of this expectation by examining over six hundred MIDs to determine whether oil ambitions motivated each militarized conflict. It finds little evidence of classic oil wars: severe militarized interstate conflicts driven largely by participants' desire to obtain petroleum resources. Instead, MIDs in oil-endowed territories were either very minor or motivated by other issues. To elaborate on these findings, I introduce four new categories of conflict: oil spats, red herrings, oil campaigns, and oil gambits. *Oil spats* are minor confrontations driven by petroleum ambitions; China and Vietnam's 2014 confrontation over a drilling rig in the South China Sea is a typical example. These episodes occur fairly regularly but never escalate into more severe interstate conflicts. The next category, *red herrings*, can be mild or severe. However, these conflicts are not fought for oil. Instead, aggressors are motivated predominantly by aspirations to political independence or regional hegemony, national security concerns, domestic politics, national pride, or contested territories' other economic, strategic, and symbolic assets. Red herrings are the dominant form of conflict in oil-endowed territories.

On rare occasions, states engage in *oil campaigns*. These major attacks target foreign petroleum resources. Yet they occur in the midst of ongoing wars that were not themselves caused by petroleum ambitions. Oil can therefore influence wars' trajectories once they are under way. But it does not start them. The only possible exception is the study's single *oil gambit*: Iraq's invasion of Kuwait. I use the term *gambit* to classify this case because it captures the conflict's instrumental character. Iraq launched an international attack targeting foreign petroleum resources in order to achieve a broader goal: subverting a perceived existential threat posed by the United States.

The book's remaining chapters conduct deeper dives into the historical conflicts that are most commonly identified as classic oil wars, framed by the four new conflict categories. Chapter 5 investigates two prominent red herrings: the Chaco War (1932–1935) and the Iran–Iraq War (1980–1988). These conflicts are widely assumed to have been oil-driven. Bolivia and Paraguay purportedly fought over the Chaco Boreal's prospective petroleum endowments, and Iraqi president Saddam Hussein supposedly invaded Iran in order to seize its oil-rich Khuzestan Province. However, a closer examination of the cases challenges these interpretations. In the Chaco War, both belligerents knew that the contested territory did not contain oil resources. In the Iran–Iraq War, Saddam's territorial ambitions were limited to small areas along the states' bilateral boundary; he repeatedly offered to withdraw from Khuzestan if those demands were met. Neither war was fought to grab petroleum resources.

Chapter 6 presents a representative oil spat between Argentina and the United Kingdom. In this 1976 incident, an Argentine destroyer intercepted the RRS *Shackleton*, a British research ship, which the Argentines believed was unilaterally exploring for oil near the contested Falkland/Malvinas Islands. The confrontation inspired intensely hostile rhetoric. However, it rapidly died down and the states began to pursue oil cooperation as a means of resolving their ongoing islands dispute. The chapter also demonstrates that the countries' later, major conflict—the Falklands War (1982)—was another red herring. Rather than being driven by oil ambitions, it was provoked by Argentine officials' determination to retake the islands before the sesquicentennial of British occupation, a rapidly changing security situation in the South Atlantic, and miscalculation of the Thatcher government's likely response.

Chapter 7 examines two prominent oil campaigns: Japan's invasion of the Dutch East Indies and northern Borneo (1941–1942) and Germany's aggression against the Soviet Union in World War II (1941–1942). As previous authors have noted, oil ambitions drove both attacks; the Japanese and Germans were desperate to acquire additional petroleum resources. However, their willingness to fight for oil was endogenous to their ongoing conflicts: the Second Sino–Japanese War (1937–1945) and World War II in Europe (1939–1945). These existing conflicts were not themselves caused by petroleum ambitions, and, in their absence, both aggressors would have refrained from fighting for foreign oil. As it was, Japan and Germany delayed their oil campaigns for as long as possible, only resorting to international aggression after alternative means of satisfying national petroleum needs had failed.

Chapter 8 evaluates the unique historical oil gambit: Iraq's invasion of Kuwait (1990). This episode tops most lists of classic oil wars, and for good reason; Saddam initiated a severe conflict that aimed to seize his neighbor's petroleum resources. Yet labeling Iraq's invasion a classic oil war is an oversimplification. Although Saddam aspired to control Kuwait's oil, his broader goal was to resist a perceived existential threat emanating from the United States. Saddam believed that the US government was inciting Persian Gulf oil producers to drive down international oil prices in order to achieve its long-standing ambitions of preventing Iraq's regional rise and removing him from power. Saddam initially tried to counter this threat domestically and diplomatically. However, after those initiatives failed, he reluctantly turned to international aggression, believing that it offered the only means of resisting the United States and possibly securing his regime's survival. Chapter 8 also includes a postscript examining the United States' 2003 invasion of Iraq, which demonstrates that even global hegemons avoid classic oil wars.

Collectively, my findings challenge the oil wars myth by revealing that states are extremely reluctant to fight for petroleum resources. Oil ambitions may inspire mild international sparring. They occasionally affect the trajectories of wars that are already under way. However, with one possible exception, oil aspirations have not caused severe interstate conflicts. Moreover, if we choose to label Iraq's invasion—the one plausible candidate—a classic oil war, we must also acknowledge that these contests look very different from the greedy petroleum grabs that we often imagine. The conclusion builds on these findings to discuss the minimal security threat posed by current and future oil competition, the reasons petroleum has not been subject to the same imperialist logic that governed states' engagement with other natural resources, the book's implications for other types of oil-related contention and US foreign policy, and the dangers of allowing myths to drive our analyses and decision making.

1

FROM VALUE TO VIOLENCE

Connecting Oil and War

"A drop of oil is worth a drop of blood." Since the early twentieth century, Georges Clémenceau's adage has regularly been deployed to support claims that countries fight over oil.[1] However, the axiom is an invention. The French prime minister did draft a hurried telegram mentioning oil to US president Woodrow Wilson in December 1917, in the midst of World War I. Recognizing that France was running short of fuel and that the United States was the world's leading petroleum producer, Clémenceau appealed to Wilson to direct American oil tankers to French shores. Yet, rather than writing, "Une goutte de pétrole vaut une goutte de sang," as the popular version of the quotation implies, the prime minister actually observed that France must have "l'essence aussi nécessaire que le sang" (the gasoline, as necessary as blood) for its upcoming battles with Germany. Otherwise, he warned, the Allied armies would be "abruptly paralyzed" and might be forced to establish an "unacceptable peace."[2] Clémenceau's telegram therefore acknowledged oil's exceptional importance for modern warfare. However, contrary to the popular misquotation, the French leader did not claim that it was worth shedding blood for oil.[3]

The cognitive leap that transformed Clémenceau's telegram from a plea for additional fuel into an explanation for violence also characterizes classic oil war claims. Rather than merely observing that petroleum is an exceptionally valuable natural resource, oil war believers also assume that petroleum is worth fighting for and, consequently, can inspire interstate conflicts. This chapter explores oil's transformation from a worthless substance into the world's most strategically and economically valuable resource and a presumed casus belli. In the process,

I elaborate on the object of my analysis—the classic oil war—and previous academic treatments of the topic. Observing that skeptical voices and inconsistent empirical evidence have failed to overturn prevailing oil war arguments, I argue that these conflicts are a myth: a taken-for-granted story about how the world works.

Oil's Value

Why would countries fight over oil? Like all natural resources, the substance has no inherent value. It is simply a collection of hydrocarbon compounds, along with small amounts of nitrogen, oxygen, sulfur, and metals, including copper, nickel, and iron. Oil can be slick or sticky, flowing easily or refusing to budge unless heated. It ranges in color from pitch black to yellowish green, and, if it possesses a high sulfur content, it smells like rotten eggs. Oil is also known as petroleum (literally, "rock oil") and, in its more viscous forms, is often identified as bitumen, tar, asphalt, or pitch.

As the resource economist Erich Zimmermann observed in the early 1950s, "Resources . . . are not, they become."[4] Oil acquired value—and its status as a "natural resource"—because of its utility to people. Mesopotamian civilizations used oil to build roads and waterproof boats. Moses's basket was caulked with pitch, and the cities of Babylon and Jericho were constructed with bitumen as mortar. Ancient Egyptians used the material in their embalming rites.[5] The Roman naturalist Pliny remarked on oil's medicinal utility; he claimed that petroleum "healed wounds, treated cataracts, provided a liniment for gout, cured aching teeth, soothed a chronic cough, relieved shortness of breath, stopped diarrhea, drew together severed muscles, and relieved both rheumatism and fever."[6] Many early Middle Eastern societies used oil as an illuminant, lighting their homes and businesses with the fuel. Some also employed oil as a weapon. In the *Iliad*, Homer reported that the Trojans launched burning bitumen against enemy ships. Over 1,500 years later, the Byzantines were renowned for their "Greek fire": incendiary petroleum applied to arrows and used in early grenades.[7]

However, it was only in the twentieth century that oil obtained its current status as a uniquely valuable natural resource. This shift arose partly from changing extractive technologies. Before the Common Era, people gathered oil solely from surface seeps. Approximately 1,700 years ago, Chinese salt miners produced the first mechanically drilled "oil" wells, which extracted natural gas, along with salt water. In the nineteenth century, prospectors began to significantly increase petroleum output by applying percussion drilling techniques to oil wells.[8] Rotary drilling, introduced around the beginning of the twentieth century, further expanded petroleum production. By 1916, oil wells exceeded 1,500 meters in depth. By the beginning of World War II, they reached 4,500 meters.[9]

Oil production also increased as the industry moved offshore. The first "offshore" well was drilled from an extended pier near Ventura, California, in 1896.[10] In the 1930s, fixed drilling platforms appeared in the shallow waters of the Gulf of Mexico and Venezuela's Lake Maracaibo. After World War II, the use of semi-submersible and floating rigs enabled offshore production to spread worldwide.[11] At the same time, the number of significant petroleum-producing countries expanded dramatically. When World War II began, only three states extracted more than fifty thousand barrels of oil per day.[12] Today, over fifty countries produce that volume. In total, the global petroleum industry's crude oil output is currently over eighty million barrels per day.[13]

The second reason for oil's changed status was shifts in petroleum consumption. Over the last 150 years, people have discovered new ways to use oil. The first major innovation was the development of kerosene. In the 1840s, Abraham Gesner, a Canadian geologist, developed a refining process to distill kerosene from bitumen.[14] The next decade, pharmacists in Galicia (in present-day Poland and Ukraine) found a means of refining the fuel from crude oil and invented a new lamp that could safely burn it.[15] Kerosene subsequently replaced whale oil and camphene as consumers' favored illuminant, boosting petroleum production.[16]

The second, more significant innovation was the development of the internal combustion engine, which ran on oil-based fuels. As the popularity of automotive transportation surged, gasoline transformed from a waste product, sold for cents on the barrel, into oil's most significant derivative.[17] Airplanes, a new innovation, also ran on petroleum-based fuels, as did growing proportions of maritime transportation. With these technological developments, the Age of Oil had begun. Petroleum production boomed to meet burgeoning demand and has continued to climb for over a century.

Oil's value, in the twentieth and twenty-first centuries, has arisen from the resource's military and economic utility. On the military side, most countries' armed services run predominantly on oil. The transition to petroleum-based fuels began in the 1880s, when British admiral John Fisher started promoting the Royal Navy's conversion from coal to oil. Fisher observed that oil-burning ships could achieve higher speeds, accelerate more quickly, and maneuver more precisely than their coal-burning counterparts. They could also be refueled at sea and required far less labor for stoking, all of which would grant the British fleet a significant advantage over its emerging naval rival, Germany. However, Fisher's initial efforts to persuade Parliament to endorse conversion floundered, as British leaders were reluctant to abandon coal, a resource they possessed in abundance, for oil, which they lacked entirely.[18] It was not until Fisher secured the support of

Winston Churchill, then First Lord of the Admiralty, that the government officially initiated the fleet's full transition to oil, in 1912.[19]

Navies saw few engagements during World War I. Nevertheless, the conflict firmly established oil's military importance. Automotive transport offered increased flexibility for modern armed forces. Rather than conducting "war by timetable" on the railways, they could use trucks to deploy troops and materiel at any time and in great volumes. Mechanized transport also revolutionized the battlefield. The invention of the tank helped armies break through the stalemate of trench warfare, while aerial reconnaissance and strategic bombing permanently altered the nature of modern war. All of these technologies ran on oil derivatives. In addition, oil-based lubricants prevented vehicles, artillery, and other military machinery from seizing up.[20]

Unsurprisingly, access to oil—and denying it to the enemy—became a major tactical concern for all of World War I's belligerents. Germany's U-boat campaign aimed to knock the United Kingdom out of the war by torpedoing fuel tankers headed toward the island nation. France was also beset by fuel shortages, prompting Clémenceau's appeal to Wilson in 1917. The United States' entry into the conflict earlier that year proved to be a crucial turning point. Despite early supply hiccups, US participation provided the Entente with sufficient oil and manpower to defeat its German adversaries, who suffered from their own fuel shortages. After the war, Britain's Lord Curzon asserted that "the Allies floated to victory on a wave of oil."[21]

Petroleum access was equally important to the outcome of World War II. Germany's dramatic victories in North Africa, from 1941 to 1942, were eventually reversed by fuel shortages; General Erwin Rommel's forces could not be resupplied because of Allied attacks on Mediterranean fuel shipments.[22] Resource shortages also contributed to the failure of Adolf Hitler's Russian campaigns. German forces lacked sufficient fuel to reach Moscow, retreat from Stalingrad, or complete Case Blue (1942), which ironically aimed to capture Soviet oil fields. Later in the war, all German operations were compromised by petroleum scarcity. By autumn 1944, much of the Luftwaffe was grounded because of inadequate fuel supplies.[23]

Oil shortages also compromised Japan's performance in the war's Pacific theater. Starting in 1944, Japanese battleships were unable to fully participate in naval engagements, including the Marianas and Philippines campaigns, because they lacked sufficient fuel supplies.[24] Oil scarcity also encouraged the Japanese to rely more heavily on their infamous kamikaze tactic, as fewer of those attacks were needed to disable an enemy battleship and no fuel was required for return flights. However, even with these brutal conservation measures, by July 1945, Japan had essentially run out of oil.[25]

Oil has sustained its military importance since the end of World War II. Contemporary military establishments continue to depend on petroleum-based fuels to power their air, sea, and land vehicles. In the United States, almost 80 percent of the energy consumption of the Department of Defense (DOD) consists of petroleum products. In 2018 alone, the US military consumed almost eighty-six million barrels of fuel for operations, training, and readiness.[26] These inputs are irreplaceable; there are currently no viable, large-scale substitutes for oil-based transportation fuels. Additionally, while the DOD has committed itself to increasing energy efficiency and renewable fuel use, these programs are still in their infancy.[27] Most other countries have made even less progress toward supply diversification. Without access to oil, every state's military would stop in its tracks.

On the economic side, oil is valuable both as an input and as a revenue source. Civilian transportation, like military transportation, runs on petroleum. In the United States, 92 percent of the transportation sector is powered by oil-based fuels.[28] Similar figures are the norm globally. In many countries, oil is used as a heating fuel or to produce electricity, in power plants or diesel-fueled generators.[29] Hydrocarbons are also employed as a feedstock for the petrochemical industry.[30] Overall, civilian oil consumption dwarfs military petroleum use. Although the DOD is the single largest oil consumer in the United States, the federal government, including the DOD, has historically accounted for less than 2 percent of US petroleum use.[31]

Losing access to civilian oil inputs can have serious negative consequences. Greater consumption of energy resources has historically been correlated with better economic performance; countries that use more energy have higher growth rates.[32] In addition, lack of access to affordable oil resources has negative economic effects; rising oil prices often precede recessions, in the United States and elsewhere.[33] Expensive oil increases household expenditures and takes a particularly heavy toll on the shipping, agricultural, and heavy industrial sectors.

Oil is also economically valuable because it can supply enormous amounts of revenue to petroleum-producing states.[34] In these countries, governments collect oil rents in a variety of ways: as concessions and royalty payments from oil companies that want to develop local petroleum resources, as taxes on oil companies' profits, and as revenue from domestic and international oil sales.[35] For many states, these revenue streams are a vital source of income. Countries like Angola, Kuwait, and Nigeria depend on oil rents for the majority of government revenue and up to 95 percent of export earnings.[36]

These governments use oil income to sustain domestic political support.[37] Some offer cash payments to citizens; in 2018, Alaska provided each resident with a $1,600 dividend from the Alaska Permanent Fund.[38] Governments also use oil rents to finance social spending, including public employment, health care and education, and to subsidize consumption of basic goods, such as food and fuel.[39] Venezuela's former president Hugo Chavez notoriously employed these strategies in the early 2000s to expand his political base.[40] During the Arab Spring (2010–2011), Saudi Arabia discouraged domestic discord by raising salaries for current government employees and devoting $134 billion to housing subsidies, unemployment benefits, and new public-sector jobs.[41] If oil-funded carrots fail to sustain public support, these governments can turn to petroleum-financed sticks; many oil producers use their resource rents to pay for robust domestic security apparatuses.[42] Gulf producers also used this strategy during the Arab Spring, spending lavishly on their own militaries and supporting their neighbors' security forces in order to contain antigovernment protests.[43]

When oil revenue declines, this political order is threatened.[44] Producer state governments may be forced to cut subsidies and reduce social spending, provoking popular discontent. After oil prices dropped in late 2014, the Saudi government eliminated bonus payments and trimmed subsidies for food and fuel. However, blowback was so severe that, in April 2017, the regime reinstated many of the benefits.[45] Similarly, the Nigerian government was compelled to restore payments to former rebels in the restive Niger Delta after a decrease in state payouts led to a resumption of hostilities.[46] Insufficient investment in domestic security institutions can also be destabilizing, as civilian and military forces have less capacity and incentive to defend the ruling regime. Countries that are highly dependent on oil rents therefore have strong inducements to defend—or increase—their petroleum revenue.

Lastly, oil producers derive some more subtle benefits from their petroleum output. One of these is a more favorable balance of trade. Major oil producers can consume domestic petroleum resources, thereby keeping their money at home rather than shipping dollars overseas to pay for foreign crude oil and petroleum products.[47] Illustratively, after the United States' shale oil boom began in 2007, the national trade deficit declined significantly.[48] Major petroleum producers may also possess political leverage over oil-consuming states, because they can threaten to shut off their petroleum supplies.[49] To stay in producers' good graces, oil-importing states provide them with significant diplomatic and military support.[50] Meanwhile, oil exporters have greater foreign policy autonomy; since they are not vulnerable to oil supply cutoffs, they do not have to tailor their behaviors to sustain their access to foreign petroleum resources.[51]

Oil is therefore an exceptionally valuable natural resource. It is a vital input for modern militaries and industrialized economies, can be a source of enormous national wealth, and conveys a variety of subtler benefits. Losing access to oil can have devastating consequences. Consequently, all countries would prefer to control more petroleum resources.[52] As Indra de Soysa and his coauthors observe,

"Countries are bound to covet assets that are valuable, tradable, conquerable, durable, and intrinsic to a given territory."[53]

Oil and Interstate Conflict?

Yet does oil's value lead countries to fight for it? Specifically, will they engage in classic oil wars: severe militarized interstate conflicts driven largely by participants' desire to obtain petroleum resources? Before answering this question, it is helpful to elaborate on the classic oil war concept. These conflicts are distinguished from other types of oil-related contention in three ways. First, classic oil wars are interstate conflicts. Unlike civil wars, which are prosecuted by actors from the same country, classic oil wars are competitions between two or more independent states.[54] Consequently, petroleum-related intrastate conflicts, such as Colombia's civil war or insurrections in the Niger Delta, are not classic oil wars.[55] Second, classic oil wars are "disproportionately lethal."[56] In these conflicts, countries expend substantial amounts of blood for oil. To reflect this criterion, my empirical analysis employs a minimum threshold of twenty-five battle deaths to identify conflicts as classic oil wars.[57]

Third, in classic oil wars, countries fight for petroleum *resources*. At least one participant aims to acquire direct, sustained control over oil or natural gas reservoirs. This characteristic distinguishes classic oil wars from other types of international petroleum-related contention, including petro-aggression, competition over oil transportation routes, and efforts to prevent the consolidation of control over global oil supplies, none of which involve seizing oil or gas reservoirs.[58] States' resource ambitions do not have to be their only motive for aggression in classic oil wars; leaders may have additional reasons for initiating these conflicts. However, in classic oil wars, the desire to obtain more oil is "a major determinant" of international attacks.[59] As Jeff Colgan asserts, in this type of conflict, "the presence or perception of oil reserves . . . creates a significant incentive for conquest."[60]

Based on these three criteria, the potential for classic oil wars is expansive. Over the last century, more than 130 countries have produced oil or natural gas, while others are believed to possess these resources.[61] Any of these states could have been the target of a classic oil war at any point after expectations of their resource endowments emerged.

Classic Oil War Believers

For countries to prosecute classic oil wars, they cannot merely recognize oil's economic and military value. Decision makers must also believe that international militarized aggression is an effective way to obtain petroleum resources and revenue. Put more simply, they must believe that fighting for oil pays, strategically or economically. This cognitive leap, from value to violence, has been endorsed by many international relations theorists. Contributors to an ongoing disciplinary debate about the value of conquest, in particular, have repeatedly asserted that fighting for oil is worth the effort.

The value of conquest debate has conventionally pitted international relations liberals and realists against each other, with the two sets of theorists disagreeing over whether foreign conquest pays in the modern world.[62] On one side, liberals assert that territorial conquest is no longer worth the effort, as changes in the international economic system, including increasing free trade, shifting patterns of global production, and land's declining value as a source of national wealth, have rendered the seizure of advanced, industrialized societies unprofitable.[63] Liberals also argue that nationalism is a serious impediment to conquering territory, as local populations resist foreign rule.[64] In addition, prospective aggressors are deterred by the difficulty of attracting foreign investment to occupied territories.[65] On the other side of the debate, realists claim that conquest can still be worth the effort if an aggressor is sufficiently repressive.[66] The disagreement over whether seizing industrialized societies pays is therefore unresolved.

Liberals and realists come together, however, on the issue of oil. Members of both camps agree that fighting for petroleum can pay. John Mearsheimer, a realist, asserts, "Any Great Power that conquered Saudi Arabia would surely reap great value from the country's petroleum resources."[67] Stephen Krasner concurs, asserting that, while the use of military force to obtain access to most raw materials is "imprudent," "in the case of petroleum, the use of force would be compatible with a politics of interest because the economic stakes are so great."[68] Kenneth Waltz and Charles Glaser are only slightly more circumspect, claiming that oil is "the only economic interest for which the United States may have to fight" and "in an era in which territory has become far less important for producing both wealth and security, territory that contains oil or controls access to it remains something of an exception."[69] Realists therefore agree that fighting for oil can be worth the effort.

Their commitment to this position is unsurprising, since realists believe, more broadly, in the value of conquest and regularly emphasize natural resources' importance as a source of state power.[70] In contrast, liberals' conviction that oil conquest pays is counterintuitive, as it departs from their usual claim that seizing foreign territory is unprofitable. Nonetheless, Christopher Fettweis states that "conquest of oil-rich regions could pay substantial dividends," while Stephen Brooks asserts that "countries with high GDP per capita whose economies are tied to extractable resources (e.g. Kuwait) still offer high cumulative gains to a conqueror."[71] Other scholars, including Daniel Deudney, Klaus Knorr, and Richard Ullman, also make an exception for oil.[72] Although resource wars rarely pay, they assert, fighting for oil can be worth the effort. As Deudney puts it, "Oil is [a] 'hard case' for the critic of resource war scenarios."[73] Liberals rationalize this departure by asserting that oil-producing countries have not undergone the economic changes that made conquering advanced, industrialized societies unprofitable. As Richard Rosecrance asserts, "Where land and its products still remain the vital factor for production—in the agriculture of Eastern Europe, the oil of the Caspian or the Middle East—territory will continue to exert a decisive influence."[74]

Few scholars, in the liberal or realist camps, offer extensive explanations for their claims that fighting for oil pays. Their petroleum-related observations are usually one-off statements or based on perfunctory reasoning. Some authors justify their assertions by highlighting oil's exceptional value. Others mention countries' intense dependence on foreign oil supplies; the lack of viable, affordable oil substitutes; or the concentration of global oil reserves in a few geographic regions. Still others base their claims on the assumption that classic oil wars have occurred in the past.[75] Yet, overall, the believers' assertions are peculiarly underdeveloped for scholars who are otherwise rigorous in their theorizing.

Classic Oil War Skeptics

Some scholars have challenged the idea that fighting for oil pays. Many argue that oil wars are inefficient. Hanns W. Maull's assertion that forcefully seizing natural resources "would be costly, of doubtful effectiveness and full of risk and uncertainties, if not even counter-productive," is representative of this view.[76] Some skeptics highlight the risks associated with specific oil grabs. Eugene Gholz and Daryl G. Press, for example, contend that seizing Saudi Arabian petroleum resources "could be accomplished only at great military, diplomatic, and moral cost."[77] Other critics claim that states refrain from fighting for oil because they can obtain the resource in other, cheaper ways: through trade or by developing petroleum substitutes. As Carl Kaysen observes, "If the calculation were made of the economic balance between securing these materials by conquest, and securing them in the ordinary ways by trade or by the search for substitutes and alternative sources of supply, it would be a peculiar situation indeed that gave the advantage to war."[78] Noting the existence of a global oil market, Brenda Shaffer describes the idea of states fighting to gain direct control over petroleum resources as "antiquated."[79]

Many skeptics argue that, for classic oil wars to occur in the contemporary international system, something would have to go terribly wrong. Gholz and Press claim that countries might seize foreign resources "in an extreme scenario."[80]

Daniel Moran and James Russell postulate that oil wars would only arise "within a context of strategic anxiety and severe economic stress."[81] Thomas McNaugher borrows from Henry Kissinger, who observed in 1974 that he would only consider seizing Middle Eastern oil if "there is some actual strangulation of the industrialized world." Facing that "gravest emergency," McNaugher asserts, the United States might fight for oil.[82]

Some scholars, however, have questioned the idea of classic oil wars entirely. Evan Luard contends that "there is no evidence that pressure on resources . . . though of acute importance to many states, especially in the case of oil, has played any part in stimulating war."[83] Other skeptics suggest that, while oil ambitions contribute to international conflict, they are a subordinate factor in leaders' war decisions. As Ronnie Lipschutz and John Holdren assert, "While resources unquestionably have played a role in the foreign and military policies of modern industrial states, this role has usually been a secondary one."[84] Similarly, Ian Lesser observes that resources have never been "the primary *cause* of war in their own right" (emphasis in original).[85]

Taken together, these skeptical voices constitute a serious challenge to oil war believers. However, individually, the critiques are piecemeal. None of the skeptics offers a comprehensive, systematic argument explaining why oil wars are unlikely. Nor have their theoretical claims been coupled with rigorous empirical analyses. Perhaps as a result, these critical voices have failed to dislodge classic oil war beliefs.

Classic Oil War Evidence

Empirical evaluations of classic oil wars are surprisingly rare, given the popular perception that petroleum competition is a serious threat to international security. Only a handful of statistical analyses have evaluated oil's contribution to international militarized conflict or territorial disputes.[86] Even fewer have produced positive findings revealing a connection between resource endowments and interstate conflict.[87] Moreover, these studies suffer from significant methodological limitations. The most severe is that statistical models do not evaluate causality; they merely search for geographic correlations between resource endowments and conflict episodes. As a result, they cannot distinguish between conflicts that are fought *for* oil resources and those that merely occur *in* oil-endowed territory. Accordingly, all statistical analyses are likely to overstate petroleum's causal power. Quantitative analyses are also unable to tell us precisely *how* oil influences international conflict, and they have difficulty parsing the intersections between petroleum resources and other causes of violence, so they capture neither the "nuance" nor the "multidimensionality" of oil-related contention.[88]

Case studies of classic oil wars do not face the same methodological limitations, as scholars can closely investigate decision makers' motives for aggression and examine conflicts within their specific historical, political, and socioeconomic contexts. However, these analyses are hampered by a different methodological weakness: their lack of generalizability. Most qualitative analyses examine only one or two supposed classic oil wars, with prominent cases, such as World War II, Iraq's invasion of Kuwait, and the Iran–Iraq War, garnering the greatest attention.[89] By focusing on these purportedly positive cases, qualitative studies imply that oil–war connections are robust. Yet, because their analytic scope is so limited, it is unclear whether these findings apply more broadly. Even if oil ambitions are a significant cause of some wars, their overall contribution to interstate conflict may be minimal.

Evidence that states engage in classic oil wars is therefore startlingly limited. Yet the conviction that countries fight to acquire oil resources persists, in spite of weak evidence, skeptical critiques, and underdeveloped supporting arguments. To understand why, it is necessary to recognize that classic oil wars are not a conventional analytic concept that can be evaluated objectively and cast aside if falsified. Instead, they are a myth.

The Oil Wars Myth

Myths, as described by international relations theorist Cynthia Weber, are "apparent truths." They are collective beliefs that are persistently reiterated until they become "that part of the story that is so familiar to us that we take it for granted."[90] Some myths emerge from historical events, while others can be traced back to religious practice. Still others arise from scientific claims about how the world works. However, over time, myths' apparent validity becomes detached from history, hard evidence, and rigorous logic. They become the ideas that we take on faith.

Like unconscious ideologies, popular discourses, and common sense, myths shape our thinking. They provide concepts, categories, and narratives that can be cast over historical and lived experience. We are more apt to notice incidents and believe in arguments that conform to existing myths. We also interpret new events and experiences through the lenses of these shared understandings.[91] By persistently attempting to make our observations fit our existing stories, we reinforce these narratives' apparent validity. Over time and through repetition, myths gain traction. They are beliefs that have become the conventional wisdom.

Since myths are "just the way things are," we rarely seek to question them. We do not investigate their empirical validity. Nor do we ask why certain myths are so widely accepted or how they unconsciously structure our understanding of how the world works, contour the way we interpret events, or condition our policy choices. We simply accept myths as accurate representations of real phenomena. As a result, topics that have been mythologized, like classic oil wars, are shielded from interrogation. We *know* they exist and assume that we understand how they work. So why bother investigating them?

I use the term *myth* to describe our collective belief in classic oil wars because it captures the way that people think about these conflicts. Classic oil wars are not simply an analytic concept developed through an objective reasoning process to reflect a real-world phenomenon and capable of being straightforwardly dismissed through empirical evaluation and critical scrutiny. Instead, they are part of our collective lore, imbued with assumptions and expectations, an enduring repository of shared fears and desires.

Classic oil wars' mythic status shields them from an array of analytic challenges and explains why scholars who are otherwise meticulous in their theoretical reasoning abandon their intellectual rigor when it comes to oil. Liberals and realists do not need to elaborate on their assertion that fighting for oil pays. Historians do not need to justify their identification of certain conflicts as oil wars. Because of the ubiquity and strength of the oil wars myth, these claims are simply presumed to be true. The myth has also contributed to the dearth of empirical analyses of classic oil wars. Since we are certain that these conflicts occur, there is little reason to devote energy to hunting for them. Finally, oil wars' mythic qualities explain the tenacity of our conviction that states fight wars to obtain petroleum resources, despite the limited historical evidence of these conflicts and substantial theoretical critiques.[92] Because myths are based on faith, they are difficult to dislodge, even with watertight logic and incontrovertible evidence.

A thorough, rigorous analysis of classic oil wars is long overdue. However, before confronting the oil wars myth, theoretically and empirically, the next chapter explores why and how oil wars were mythologized. What is it about this particular type of conflict that captures the popular imagination, making classic oil wars appear exceptionally believable?

2

EXPLAINING THE OIL WARS MYTH

Mad Max and El Dorado

Science must begin with myths, and with the criticism of myths.

—Karl Popper

Why is the oil wars myth so widely accepted? Why are people so ready to believe that countries fight over petroleum resources? One reason is the attraction of simple explanations. By reducing the causes of conflict to one factor—oil—the myth provides a parsimonious explanation for international violence. A second reason is petroleum's exceptional value. Given the resource's unique military and economic utility, the idea that countries fight to obtain it seems eminently plausible. Nevertheless, this chapter argues that the potency of the oil wars myth does not rest solely on theoretical parsimony or petroleum's value. Oil has become "a critical component of intuitive explanations about the causes of war in the modern world" because the idea of countries fighting for petroleum aligns with two hegemonic myths about the causes of resource-related conflict: Mad Max and El Dorado.[1]

These two hegemonic myths posit different motives for violence. The Mad Max myth proposes that actors fight out of need. According to this narrative, individuals, groups, and countries are locked in existential struggles and must acquire certain, vital materials in order to survive. They fight for these materials because the consequence of failing to obtain them is death. The El Dorado myth, in contrast, asserts that actors fight out of greed. They aspire to grab copious amounts of valuable materials in order to increase their wealth. They engage in violence because it is profitable.[2]

Although the labels are new, these two hegemonic myths have permeated academic and popular discourses for centuries. The Mad Max myth emerged with Malthusian arguments about the consequences of unchecked population growth. The El Dorado myth was crafted by Spanish conquistadors, although its roots can be traced back to the classical age. Since they originated, the myths have persistently reappeared in scholarly and popular representations of the causes of interpersonal, intergroup, and international resource-related conflict. The myths' power therefore arises in part from their familiarity. Repetition, in numerous settings, has reinforced their credibility. Over time, the Mad Max and El Dorado myths became accepted understandings of how the world works. They are now the conventional wisdom.

The myths' power also arises from their narrative structure. Although the Mad Max and El Dorado myths highlight specific motives for violence—need and greed—they offer much more than one-word explanations. They tell stories about how the struggle for existence and the pursuit of wealth lead to violent conflict. By tracing apparently credible pathways from motives to outcomes, the myths draw us in, encouraging us to overlook any false assumptions or logical inconsistencies in their storylines.[3] The Mad Max and El Dorado myths are easy to believe.

Each myth offers a compelling explanation for classic oil wars. According to the Mad Max myth, countries fight for oil because they need it for national survival. Alternatively, according to the El Dorado myth, they fight for oil because they greedily aspire to increase their national wealth. The narratives therefore provide two compelling foundations for the oil wars myth. They are, to paraphrase Cynthia Weber, what makes the classic oil war story make so much sense.[4] Moreover, by providing two distinct but complementary stories, the narratives render classic oil wars exceptionally believable. If El Dorado fails to rationalize a given conflict, Mad Max can step in. Thus, regardless of a person's foundational beliefs about actors' motives for violence, classic oil wars appear to be plausible events. When people accept both narratives, the oil wars myth is doubly credible.

In tracing the intellectual histories of the Mad Max and El Dorado myths, I do not seek to discredit them; I am agnostic about their accuracy as representations of actors' motives for violence. Instead, I aim to bring the oil wars myth's foundations out of the shadows of taken-for-granted knowledge and expose them to the light of critical scrutiny. These hegemonic narratives, like the oil wars myth, should be subject to question rather than accepted on blind faith. By revealing their persistence and showing how they structure popular understandings, including belief in classic oil wars, the chapter breaks their unconscious hold on our thinking, enabling us to challenge the oil wars myth itself.

Need and Greed

Scholars have identified need and greed as central motives for violent conflict since at least the seventeenth century. In his classic work *Leviathan* (1651), the English philosopher Thomas Hobbes identified three reasons that people fight each other: diffidence, competition, and glory.[5] The first two of these motives parallel the contemporary concepts of need and greed. "Diffidence," as presented by Hobbes, is existential insecurity. In an anarchic world, without a central authority to protect them, people fear for their survival. Although they would prefer to avoid violent conflict, they must sometimes fight to defend themselves and the "modest" goods they require to live. Individuals motivated by diffidence therefore "invade . . . for safety"—that is, because of need. The second motive, competition, is effectively greed. According to Hobbes, individuals driven by this motive "invade for gain." Unsatisfied with the basic goods required for their survival, they "use violence, to make themselves masters of other men's persons, wives, children, and chattel." They undertake "acts of conquest, which they pursue farther than their security requires."[6]

Later international relations theorists developed their own need–greed dichotomies to classify actors' motives for conflict. Hans Morgenthau asserted that people fight because of either conflicts of interest or their drive to dominate (*animus dominandi*). In conflicts of interest, individuals are concerned with their "vital needs." They may nonetheless engage in violence when two actors require the same thing, which only one can possess; under these zero-sum circumstances, Morgenthau observed, "struggle and competition ensue." In contrast, individuals motivated by the *animus dominandi* employ violence to obtain more power than they require for their survival.[7] Charles Glaser offers a similar dichotomy, distinguishing between "not-greedy" and "greedy" states. Not-greedy states, he claims, pursue their own survival. They have limited material ambitions but may still fight if their security depends on it. Greedy states, in contrast, are "willing to incur costs or risks for nonsecurity expansion."[8] They attempt to obtain more than they need to survive.

The need–greed dichotomy is even more prominent in civil war studies. Since the 1990s, researchers have characterized rebels as needy or greedy, depending on their primary motive for aggression.[9] Needy rebels are motivated by grievances; they fight to rectify political, ethnic, or material inequalities, including maldistribution of natural resource rents. These combatants' goal is to obtain "the general qualities required by people for their existence."[10] Greedy rebels, in contrast, are motivated by a desire for gain. They challenge the central government or seek greater autonomy in order to grab natural resource rents and enrich themselves.[11] They are pursuing more than survival.

All of the need–greed dichotomies are simplifications. Both motives are ideal types and may be difficult to distinguish in practice; the line between actors that merely seek sufficient materials for their survival and those that aspire to a surplus is blurry. Nonetheless, together, these two ideal types offer a complementary pair of rationales for resource-related violence. Actors may fight for valuable raw materials because they require them to survive. Or they may fight for resources in order to amass great wealth. Both of these motives offer plausible explanations for resource wars.

However, single-word explanations, alone, rarely capture the popular imagination, even when conveyed with a punchy rhyme scheme. The need and greed motives have been embraced because they form the cores of two hegemonic myths: Mad Max and El Dorado. These myths, which have existed for centuries, tell the stories of how need and greed lead to violent conflict. It is the availability of these easily accessible narratives that renders both motives particularly plausible. Mad Max and El Dorado are the reasons that we believe in resource-related violence, including classic oil wars.

The Mad Max Myth

The Mad Max myth is named for the cult film series, which was launched with the eponymous *Mad Max* in 1979.[12] The films take place in a postapocalyptic landscape where the remaining human population is fighting for its survival. In the series' second installment, *Mad Max 2: The Road Warrior* (1981), a narrator explains how "this wasted land" came to be. "We have to go back to the other time," he intones, "when the world was powered by the black fuel and the desert sprouted great cities of pipe and steel." The narrator explains that this oil-powered world was brought down by a great war. "For reasons long forgotten," he pronounces, as the screen flickers through stock footage of twentieth-century conflicts, "two mighty warrior tribes went to war and touched off a blaze, which engulfed them all."[13]

The war shuttered oil production, leading to the collapse of industrialized societies. "Without fuel they were nothing," the narrator expounds. "They'd built a house of straw. The thundering machines sputtered and stopped. . . . Their world crumbled." The consequences of the petroleum shortage were devastating: "Cities exploded: a whirlwind of looting, a firestorm of fear. Men began to feed on men. On the roads it was a white-line nightmare. Only those mobile enough to scavenge, brutal enough to pillage would survive. The gangs took over the highways, ready to wage war for a tank of juice."[14] The rest of the film depicts a zero-sum battle over gasoline. A group of plucky civilians that controls an oil well and refinery confronts a biker gang that is determined to seize those facilities. The ensuing clashes are presented in stark, existential terms. Those who obtain fuel may live. Those who do not will die. The civilians eventually prevail, with Mad Max's help. However, it is clear that the fight for vital resources will continue.

The Mad Max franchise did not invent the idea that competition over critical resources provokes violent conflict. Rather, the films are particularly vivid instantiations of a hegemonic myth that has existed for centuries. I label it the Mad Max myth partly because the films have become a cultural shorthand for describing this type of desperate, dystopian, zero-sum struggle, but also because identifying the myth with a fictional narrative reminds us of its possibly illusory qualities. The Mad Max myth may reflect reality. But alternatively, it may not.

The Mad Max myth has appeared in many scholarly and popular guises. In all of these instances, it presents a consistent storyline. Actors—whether they are individuals, groups, or countries—are engaged in a struggle for existence. They require certain natural resources to survive. Yet, as a result of overconsumption or degradation, available resource supplies are insufficient to meet everyone's needs. Scarcity of these critical materials drives actors to desperate measures, including violent conflict. In short, actors fight because they need resources to live.

The modern progenitor of the Mad Max myth was the English theologian and economist Thomas Robert Malthus (1766–1834). In his *Essay on the Principle of Population* (1798), Malthus issued a gloomy prediction. Because population increases geometrically, while food production increases only arithmetically, societies will inevitably run short of sustenance. Food scarcity produces want and illness, at best, and famine and death, at worst. Malthus asserted that this dynamic was inescapable: a law of nature. All societies were condemned to periodic "misery and vice" brought on by food shortages.[15]

One of these vices was war. Malthus observed that, as societies grew, the search for sustenance compelled some members to expand into new territories. As he put it, "Young scions were then pushed out . . . and instructed to explore fresh regions and to gain happier seats for themselves by their swords." If the lands they moved into were empty, shortages would be temporarily resolved. However, if the lands were already occupied, this emigration would trigger violent conflicts. "When they fell in with tribes like their own," Malthus claimed, "the contest was a struggle for existence." These zero-sum conflicts were inevitably intense. The losers would be exterminated by the victors or, deprived of food supplies, would perish "by hardship and famine." Hence, groups "fought with a desperate courage, inspired by the reflection that death was the punishment of defeat and life the prize of victory."[16] In the aftermath of these existential battles, the victors' food needs would be temporarily sated. However, Malthus observed grimly, there was no escaping a law of nature. Eventually, the victorious population would again outstrip its food supplies, prompting further want, expansion, and war.

In later editions of his work, Malthus provided extensive examples of groups fighting over scarce food resources. He drew on contemporary anthropological research, including chronicles of Captain James Cook's voyages, to describe patterns of violent conflict among New Zealanders, South Pacific Islanders, and Native Americans. He claimed that these societies were frequently confronted with food scarcity and "it may be imagined that the distress must be dreadful." Quoting Cook, Malthus surmised that the groups were "perpetually destroying each other by violence, as the only alternative of perishing by hunger."[17] Malthus suggested that these contests had biblical precedents and attributed persistent tribal warfare in Arabia and Central Asia to shortages of fertile agricultural land and pastures. He also applied his theory to Europe, asserting that sustenance needs had driven expansion and violence under the Roman Empire and during the Dark Ages.[18]

Malthus's *Essay* was widely read during his lifetime and captured the imaginations of many later thinkers. One of the first to acknowledge his indebtedness to Malthus was Charles Darwin. In his *Autobiography* (1887), the naturalist wrote that he read the *Essay* in autumn 1838.[19] By that point, Darwin had already developed the idea of natural selection. However, he was missing a key piece of the puzzle: a rationale for why the process occurs. Malthus's concept of a struggle for existence, brought on by the imbalance between population and food supplies, provided that underlying stimulus. As Darwin later wrote, "Reading Malthus, I saw at once how to apply this principle."[20] Scarcity prompted competition, which the fittest individuals survived, while the less fit perished.

In an 1844 essay, Darwin drew an explicit connection between his work and the theologian's, claiming that his evolutionary theory was "the doctrine of Malthus applied in most cases with ten-fold force."[21] Darwin also referred to the intellectual inheritance in his most famous work, *On the Origin of Species* (1859). He reiterated that his theory was "the doctrine of Malthus, applied with manifold force to the whole animal and vegetable kingdoms," and presented his argument in Malthusian terms. "More individuals of each species are born than can possibly survive," the naturalist explained; "consequently, there is a frequently recurring struggle for existence."[22]

Initially, Darwin only applied Malthus's logic to the plant and animal kingdoms. These nonhuman species, Darwin argued, lacked the "moral restraint" that could act as a check on population growth, thereby mitigating the struggle for existence among humans.[23] However, in *The Descent of Man* (1871) and in his personal correspondence, Darwin flirted with applying his theory to people. In *Descent*, he asserted that, "as man suffers from the same physical evils as the lower animals, he has no right to expect an immunity from the evils consequent on the struggle for existence."[24] Other authors also embraced this social extension of Darwin's argument. Herbert Spencer, who famously coined the phrase "survival of the fittest," claimed that the struggle for existence and natural selection were universal mechanisms. They operated within human societies, he asserted, as well as among nonhuman species.[25]

By the end of the nineteenth century, political geographers were applying Malthusian and Darwinian arguments to nation-states.[26] Friedrich Ratzel, the "father of political geography," conceived of states as organisms and argued that, like plant, animal, and human populations, they have an inherent tendency to expand. For states, expansion meant acquiring additional territory, which he referred to as lebensraum (living space). Ratzel asserted that, like other organisms, states were subject to natural selection. Accordingly, in the course of their expansion, stronger states would naturally displace weaker ones. This "struggle for space" was the geopolitical equivalent of Darwin's "struggle for existence."[27]

Political geographers argued that, to increase their chances of survival, states expanded into areas that were strategically and economically valuable. One of their aims was to satisfy a Malthusian compulsion to acquire sufficient food supplies. However, states also needed to obtain critical raw materials, such as iron and coal, which were increasingly important contributors to countries' economic and political survival.[28] These expansionist efforts were likely to provoke interstate conflicts. By the late nineteenth century, when the discipline of political geography emerged, the world had become a very crowded place. Europe was fully divided into contiguous nation-states, and the preceding century's rush for colonies had apportioned much of the rest of the globe. Consequently, countries that needed more land or vital resources could not simply expand into unclaimed territories. Instead, as Vladimir Lenin observed in *Imperialism* (1917), they would have to displace each other, triggering interstate conflict.[29]

Geopolitical arguments, emphasizing the need for secure access to vital raw materials, flourished during the interwar period. However, they fell out of favor after World War II, because of their association with Nazi Germany.[30] Nevertheless, similar arguments revived only a quarter century later. In 1972, Nazli Choucri and Robert North introduced "lateral pressure theory," which argued that "growing population and developing technology places rapidly increasing *demands* upon resources, often resulting in internally generated *pressures*" (emphasis in original). This pressure prompted states to expand and, when multiple countries pursued the same strategy simultaneously, provoked international violence. As Choucri and North asserted, "There is a strong possibility that eventually the two opposing spheres of interest will intersect. The more intense the *intersections*, the greater will be the likelihood that competition will assume *military* dimensions. When this happens, we may expect competition to become transformed into conflict" (emphasis in original).[31] The authors attributed a number of interstate conflicts, including World War I, to lateral pressure.[32]

The 1970s were a welcoming environment for neo-Malthusian arguments like Choucri and North's. In the late 1960s, a significant increase in world population propelled books like Paul Ehrlich's *Population Bomb* (1968), which highlighted the dangers posed by demographic pressure, onto the best-seller list.[33] *The Limits to Growth* (1972), a study commissioned by the Club of Rome, provoked similar fears of natural resource shortages.[34] Petroleum scarcity, in particular, was a growing concern. In 1972, the former US secretary of the interior Stewart Udall warned that declining US oil production would soon lead to petroleum shortages.[35] Six months later, James E. Akins, the director of the US State Department's Office of Fuels and Energy, echoed Udall's fears in a *Foreign Affairs* article entitled "The Oil Crisis: This Time, the Wolf Is Here."[36]

These pessimistic predictions proved to be prescient. By summer 1973, US consumers were facing gasoline shortages, precipitated partly by resource scarcity, but also by the Nixon administration's oil price controls. The energy crisis intensified that October, when Arab members of OPEC (the Organization of the Petroleum Exporting Countries) raised the price of oil, cut petroleum production, and restricted exports to the United States and several other countries in retaliation for those countries' support of Israel in the 1973 Arab–Israeli War. The US government imposed rationing programs on gasoline and heating oil, gas stations ran out of fuel, and Americans queued for hours to fill up their tanks. Truckers staged violent blockades to protest high gasoline prices, and fights broke out at gas stations as motorists attempted to obtain needed fuel.[37] The Iranian Revolution, in 1978–1979, precipitated a second energy crisis. Recalling their earlier experience with fuel scarcity, Americans engaged in panic buying, exacerbating gasoline shortages and provoking violence. Truckers again protested, sparking riots, and, in May 1979, a motorist was shot and killed in a gas line.[38]

The first two Mad Max films were released in 1979 and 1981, in the midst of the second energy crisis. Although produced in Australia, rather than the United States, they reflected the intense popular anxiety about resource shortages that prevailed in many industrialized countries at the time. Under these disquieting conditions, it was easy to believe that people and countries would fight over scarce, vital resources, especially oil. The Mad Max myth's credibility was further buttressed by politicians, who suggested that petroleum shortages could lead to violent conflicts. In late 1974, then–US secretary of state Henry Kissinger asserted that the United States might attempt to seize foreign oil if the industrialized world faced "strangulation" by Arab producers.[39] In January 1980, President Jimmy Carter issued the Carter Doctrine, asserting that the United States would respond forcefully to any attempt to interrupt Persian Gulf oil flows. Ronald Reagan issued warnings about impending resource wars during his 1980 presidential campaign, and, in autumn 1981, the new US secretary of state, Alexander Haig, boldly announced that "the era of the resource war has arrived."[40] Alarmist scholarship on the subject flourished.[41]

By the mid-1980s, however, these fears had subsided. The oil price collapse in 1986, following an improvement in access to mineral supplies from southern Africa and a declining Soviet threat to the Persian Gulf, temporarily curtailed resource scarcity concerns.[42] Yet Malthusian narratives revived, in a different guise, only a few years later. In the 1990s, numerous research programs, including Thomas Homer-Dixon's Toronto Group, began to argue that scarcity of critical renewable resources, including cropland, timber, and water supplies, could inspire intrastate conflict.[43] The journalist Robert Kaplan popularized these arguments in a notorious *Atlantic Monthly* article entitled "The Coming Anarchy." Predicting that population growth and resource competition would inevitably lead to disease, poverty, and violence, Kaplan returned the Mad Max myth to its intellectual roots. "It is Thomas Malthus," he asserted, "the philosopher of demographic doomsday, who is now the prophet."[44]

Kaplan's alarmist claims found a sympathetic political audience. President Bill Clinton was so stirred by "The Coming Anarchy" that he faxed it to all American embassies.[45] "I was so gripped by many things that were in that article and by the more academic treatment of the same subject by Professor Homer Dixon," he later claimed in a speech to the National Academy of Sciences. The president also explicitly linked these articles to earlier Malthusian narratives; "You could visualize a world in which [we] . . . look like we're in one of those Mel Gibson 'Road Warrior' movies," he stated.[46] Kaplan's article also purportedly inspired Vice President Al Gore to create the State Failure Task Force to analyze the connections between environmental degradation and state collapse.[47] People found the idea of subnational conflicts over critical resources eminently plausible.

By the 2000s, the specter of interstate resource conflicts had also revived. In his popular book *Rising Powers, Shrinking Planet*, Michael Klare argued that growth in natural resource demand, coupled with declining global supplies, would "inevitably" lead to international violence. He described this threatening dynamic in Darwinian terms. States were "predators" that were "hungry" and "thirst[y]" for vital natural resources. To satisfy these needs, they engaged in "a ferocious struggle over diminishing sources of supply."[48] His extended description of these conflicts mirrored Malthus's: "Those that retain access to adequate supplies of critical materials will flourish, while those unable to do so will experience hardship and decline. The competition among the various powers, therefore, will be ruthless, unrelenting, and severe. Every key player in the race for what's left will do whatever it can to advance its own position, while striving without mercy to eliminate or subdue all the others."[49] The consequences of resource competition could be devastating; Klare warned that "the potential to slide across this threshold into armed conflict and possibly Great Power confrontation poses one of the greatest dangers facing the planet today."[50]

Much of Klare's work has emphasized the risks of oil wars specifically. In *Rising Powers, Shrinking Planet*, he presented international competition over petroleum resources as "a voracious, zero-sum contest that, if allowed to continue along present paths, can only lead to conflict among the major powers."[51] Klare also claimed that these conflicts had significant historical precedents; as he put it, "Governments have repeatedly gone to war over what they view as 'vital national interests,' including oil and water supplies."[52] As "peak oil" fears intensified in the mid-2000s, many authors issued similarly alarmist claims about the prospects of future petroleum conflicts, especially between the United States and China. Great powers were expected to fight over the world's diminishing oil supplies.[53]

The Mad Max myth permeates popular culture, as well as academic and political discourses. Many films depict competitions over scarce, vital resources. In classic Westerns, water conflicts are a recurrent trope. John Wayne fights for water access in *Riders of Destiny* (1933) and *King of the Pecos* (1936).[54] In *The Big Country* (1958), Gregory Peck becomes enmeshed in a struggle over cattle watering rights.[55] Water conflicts also appear in *Chinatown* (1974), which fictionalizes the historical struggle over the Owens Valley water supply, and *The Milagro Beanfield War* (1988), in which local farmers' efforts to water their fields spark contention with land developers.[56] The heroes of *Tank Girl* (1995), the James Bond film *Quantum of Solace* (2008), and the fourth installment of the Mad Max series, *Mad Max: Fury Road* (2015), also fight against powerful adversaries who have monopolized control over local water resources.[57] In all of these films, the protagonists need water to survive and can only obtain it through violence.[58]

Resource scarcity and its noxious effects are also prominent features of many postapocalyptic and disaster movies. The premise of films like *Avatar* (2009), *Elysium* (2013), *Interstellar* (2014), and *Wall-E* (2008) is that overpopulation, overconsumption of natural resources, and natural disasters have pushed humankind to the brink of collapse.[59] In some of these films, the remaining human population responds to resource shortages by migrating: relocating to space stations or other planets. In others, societies have developed more novel—and unpleasant—ways to cope with scarcity. In *Logan's Run* (1976), people are executed at the age of thirty to conserve resources. In *Soylent Green* (1973), the population is fed with human remains.[60]

Resource scarcity also prompts violent conflicts. In many postapocalyptic film and television landscapes, including those of *The Walking Dead* (2010–) and *Into the Badlands*, (2015–2019), people fight each other for vital resources.[61] Many alien movies depict Malthusian struggles on an interplanetary scale; extraterrestrials descend on Earth in order to secure access to critical materials. In *The War of the Worlds* (1953), the Martians invade because of water, clean air, and natural resource shortages on their home planet. As the film's narrator explains, "Mars is more than 140 million miles from the sun, and for centuries has been in the last status of exhaustion. . . . Inhabitants of this dying planet looked across space with instruments and intelligences that which we have scarcely dreamed, searching for another world to which they could migrate."[62] Aliens' efforts to obtain needed resources—usually by exterminating humans—also drive aggression in *Independence Day* (1996), *Battle: Los Angeles* (2011), and *Oblivion* (2013).[63]

Need-driven conflicts over oil, specifically, are a relatively uncommon plot device. Although politicians and academics regularly present oil as a vital state need, culturally, oil competition is more commonly associated with greed. However, in addition to propelling *The Road Warrior*, need-driven oil conflict is the linchpin of the television series, *Occupied* (2015–).[64] In this series' fictionalized world, Norway is a critical petroleum supplier, since wars have interrupted Middle Eastern oil production and the United States has halted petroleum exports. However, the environmentally minded Norwegian government has decided to shutter its oil industry in order to combat climate change. The European Union, desperate for fuel, endorses a Russian invasion of the recalcitrant Scandinavian producer. Although initially peaceful, the occupation soon sparks local protests, armed resistance, and military crackdowns. The logic underpinning this violence is never questioned; clearly, the series implies, countries will fight for oil when they need it for their survival.

The same logic appears at the end of Sydney Pollack's thriller *Three Days of the Condor* (1975). Having discovered that a renegade CIA unit killed his colleagues to conceal its plan to seize Middle Eastern oil, Turner (Robert Redford) confronts Higgins (Cliff Robertson), insisting to his superior that the American people would be appalled by the agency's behavior. Higgins responds scornfully, "Ask them when they're running out. Ask them when there's no heat in their homes and they're cold. Ask them when their engines stop. Ask them when people who have never known hunger start going hungry. You want to know something? They won't want us to ask them. They'll just want us to get it for them!" When faced with an existential crisis, people will embrace international oil grabs.[65]

Mad Max therefore offers one explanation for classic oil wars. According to this narrative, petroleum provokes violent conflict when countries need more oil in order to survive. The narrative is compelling, both because of oil's exceptional military and economic value and because of the apparent plausibility of the Mad Max myth, whose grounding in Malthusian and Darwinian arguments gives it a veneer of natural law, while its repetition, in numerous scholarly and popular contexts, makes it an easily accessible trope. Mad Max offers a consistent, resonant story about how actors respond to natural resource scarcity. Nonetheless, the intensity of classic oil war convictions cannot be attributed only to the Mad Max myth. The El Dorado myth offers an equally compelling narrative, explaining why countries fight over oil. This story, too, is grounded in natural resources' value. However, it emphasizes resources' ability to convey enormous wealth rather than their capacity to satisfy existential imperatives. In the El Dorado myth, it is greed, not need, that drives violence.

The El Dorado Myth

> *"Over the Mountains*
> *Of the Moon,*
> *Down the Valley of the Shadow,*
> *Ride, boldly ride,"*
> *The shade replied,*
> *"If you seek for Eldorado!"*
> —Edgar Allan Poe, "Eldorado"

El Dorado was originally a mythical individual. In the 1530s, rumors began to circulate among Spanish conquistadors about a gilded man, living in the Colombian highlands. The story's initial source was reportedly an indigenous guide from Quito who recounted his tale to the followers of Sebastián de Benalcázar. The guide described an unusual ritual practiced by the Chibcha group, from Cundinamarca Province, near present-day Bogotá.[66] He claimed that, when the Chibcha appointed a new leader, they ritually anointed the man with a sticky resin and then covered him in gold dust. The leader was rowed out to the middle of Lake Guatavitá, where he washed off the gold dust and cast other gilded artifacts into the water.[67]

The conquistadors were a receptive audience for stories of golden treasure. Hernán Cortés's 1519 seizure of Tenochtitlan, the capital of the Aztec Empire, which garnered him astronomical amounts of gold, had already become the stuff of legend. Benalcázar's men may also have heard tales of Francisco Pizarro's 1532 defeat of the Inca Empire. The captured Inca leader, Atahualpa, had offered Pizarro a room filled once with gold and twice with silver in exchange for his release. To the conquistadors' delight, Atahualpa fulfilled his side of the bargain. The Spaniards, however, reneged on their promise. Having obtained the Incas' gold, they executed their leader—one of the many acts of violence perpetrated during the Spaniards' search for gilded treasure.

The apparent abundance of Latin American gold resources and the ease with which Cortés and Pizarro had seized them suggested that further riches could be ripe for the taking. The tale of El Dorado, which provided a hint as to where the next fortune might lie, triggered an impassioned hunt that would last for almost a century. As Charles Nicholl writes, "The idea of El Dorado: the probability that it was there, the possibility of finding it, the untold riches it contained—was a craze that gripped people. It had the force field of a cultish religion."[68] In 1536, Gonzalo Jiménez de Quesada hunted for the gilded man and his treasure in the Colombian interior. His brother, Hernán Pérez de Quesada, tried to drain the fabled Lake Guatavitá in order to collect golden artifacts. In 1540, Gonzalo Pizarro, Francisco's half-brother, led an expedition into the Amazon Basin, searching for El Dorado.[69] In addition, a number of German conquistadors, including Georg von Speyer, Nikolaus Federmann, and Abrosius Ehringer, hunted for El Dorado in Venezuela and Colombia from the mid-1530s to the 1540s.[70]

Over time, El Dorado evolved from a person into a place. In the 1590s, Sir Walter Raleigh scoured Guyana not for a gilded leader but for a city whose buildings and streets were paved with gold. Raleigh claimed to have discovered the city in 1595.[71] However, his expedition, like those that preceded it, was largely unsuccessful. None of the treasure hunts unearthed extensive riches, and scores of people perished along the way. Often, searches for El Dorado devolved into ruthless violence, as conquistadors and their retinues fought among themselves and with local populations. Contemporary descriptions of Ehringer's expedition are particularly vivid, reporting that "chieftans were enslaved . . . rebellious captives were burned alive, even friendly Indians, bringing gifts, were cut to pieces."[72]

The contours of the El Dorado myth are therefore consistent, even though the legend's precise target and location have shifted over time. At the myth's core is the idea of fabulous wealth. The riches that exist, either in the gilded man's body or in a golden city, are assumed to be dazzlingly large. Anyone who finds them will obtain wealth beyond his wildest dreams and a lifetime of comfort and pleasure. People's feverish desire to obtain these riches provokes intense greed, which can inspire extreme acts, including violence.[73]

The El Dorado myth is intertwined with a number of other, similar legends. Francisco Coronado spent years crisscrossing Mexico and the American Southwest in search of the legendary Seven Cities of Cibola, a group of settlements that were supposedly encrusted with jewels and gold.[74] Another conquistador, Juan Ponce de Léon, traversed the Florida cays, hunting for the fountain of youth. In Greek mythology, Jason and his Argonauts pursued the Golden Fleece, and in Arthurian lore, the knights of the Round Table sought the Holy Grail.[75] Each of these searches was driven by desire for an enormously valuable treasure. And each group that sought these fabled artifacts faced privations, danger, and violence during its quest.

The El Dorado myth has not been as prominent in scholarly discourses as the Mad Max myth. There are no historical equivalents to Malthus, Darwin, or Ratzel that offer a general theory of how actors' resource greed provokes violent conflict. Contemporary civil wars scholarship, however, is replete with accounts of greedy rebels enriching themselves by prosecuting intrastate conflicts in gold-, diamond-, and oil-endowed territories.[76] At the international level, critiques of imperialism often possess El Dorado undertones. Jack Snyder, for example, identifies "El Dorado and Manifest Destiny" as one of the myths that fuel imperial overexpansion.[77] In addition, researchers often attribute specific acts of international aggression to resource greed; Saddam Hussein's invasion of Kuwait, in particular, is regularly depicted in these terms.[78]

Nonetheless, the El Dorado myth—and the themes that underpin it—have been more prominent in literature and popular culture than in international relations theory. Narratives about gold's pernicious effects date back to the classical era. In his epic poem *Metamorphoses* (8 AD), Ovid blamed the mineral for the fall of man:

> Thus cursed steel, and more accursed gold
> Gave mischief birth, and made that mischief bold;
> And double death did wretched man invade,
> By steel assaulted, and by gold betrayed.[79]

Because of desire for gold, Ovid asserted, "mankind is broken loose from moral bands." Interactions that were once characterized by "truth, modesty, and shame" are now dominated by "fraud, avarice, and force."[80] In the *Aeneid* (29–19 BC), Virgil presented several stories of men's gold lust provoking terrible, violent behaviors. Pygmalion murdered his sister Dido's husband in order to obtain great wealth; as Virgil chronicled, "Then strife ensued and cursed gold the cause."[81] Polymnestor murdered Polydorus, King Priam's son, after accepting a gilded payment to protect him. "O sacred hunger of pernicious gold!" the poet lamented, "What bands of faith can impious lucre hold?"[82]

In *The Divine Comedy* (1320), Dante Alighieri, who drew on the *Aeneid* for inspiration, placed Pygmalion and Polymnestor in the fourth circle of hell, where biblical, mythical, and historical individuals are punished for greed. Another of the condemned, Achen, was censured for stealing gold and silver from the spoils of the Battle of Jericho, which provoked God to cause the Israelites' defeat at Canaan.[83] Dante also expounded on the dangers posed by greed in his earlier work *Convivio*. "What else," he queried, "imperils and slays cities, countries, and single persons so much as the new amassing of wealth by anyone?" The goal of acquiring riches, Dante cautioned, "may not be reached without wrong to someone."[84]

Moreover, he warned, avarice is never satisfied. Although "the false traitoresses promise . . . to remove every thirst and every want and to bring satiety and sufficiency," in practice, obtaining some wealth only produces more "feverish" desire. "For never," the poet observed, "is the thirst of cupidity filled nor sated."[85]

Gold lust also inspired violence in one of the stories in Chaucer's fourteenth-century *Canterbury Tales*. "The Pardoner's Tale" depicts three young men who set out to find and kill Death. Following an old man's instructions, they search for Death under a particular oak tree. When they arrive at the specified locale, they see only a large store of gold coins. The young men joyfully plan to divide their new riches and travel onward the next day. However, during the night, their greed for gold inspires violent betrayals. Two of the men kill the third to obtain a greater share of the wealth. However, the murdered man has already poisoned the others' wine. Thus, by morning, all three have found Death.[86]

Contemporary retellings of "The Pardoner's Tale" have framed the story as a golden treasure hunt. In *The Treasure of the Sierra Madre* (1948), three prospectors, played by Humphrey Bogart, Tim Holt, and Walter Huston, decide to search for gold in Mexico's Sierra Madre mountains. The miners expect to acquire exorbitant riches. As Bogart's character, Fred Dobbs, claims, "This is the country where the nuggets of gold are just cryin' for ya to take 'em out of the ground and make 'em shine." He optimistically predicts that, "If we make a find, we'll be lightin' our cigars with hundred-dollar bills."[87]

The miners' quest is initially successful; they discover a rich vein of gold and extract enough ore to make them all extremely wealthy. However, the appearance of gilded treasure sparks intense greed. As Huston's character, a grizzled old prospector named Howard, observes early in the film, "When the piles of gold begin to grow . . . that's when the trouble starts." Howard recognizes that gold lust is insatiable. As he says to the other miners, "I tell you, if you was to make a real strike, you couldn't be dragged away. Not even the threat of miserable death would keep you from trying to add ten thousand more. Ten you want to get twenty-five. Twenty-five you want to get fifty. Fifty, a hundred. Like roulette. One more turn, you know. Always one more."[88]

As Howard predicts, Dobbs's insatiable gold lust eventually drives him to violence. Desiring all the riches for himself and fearing that his partners plan to steal his share, Dobbs shoots Holt's character and grabs everyone's gold. While trying to escape, however, he is captured and killed by bandits. Describing Dobbs's noxious acts, Howard is empathetic. Any man could have given into the temptation, he claims, including himself. The pursuit of El Dorado can incite anyone to violence.

Many other popular films have presented El Dorado narratives. An animated feature, *The Road to El Dorado* (2000), portrays a hunt for the gilded city.[89] Harrison Ford searches for El Dorado in the final installment of the Indiana Jones series, *Indiana Jones and the Kingdom of the Crystal Skull* (2008). The series' third film, *Indiana Jones and the Last Crusade* (1989), features a hunt for a golden cross that belonged to Francisco Coronado, as well as a quest for the Holy Grail.[90] In *National Treasure: Book of Secrets* (2007), Nicolas Cage searches for Cibola.[91] Other films, including *Treasure Island* (1950), *The Good, the Bad, and the Ugly* (1967), *Three Kings* (1999), *Goonies* (1985), and *The Pirates of the Caribbean: The Curse of the Black Pearl* (2003), also depict golden treasure hunts.[92] All of movies involve at least the threat of violence, as actors pursue enormous wealth.

The El Dorado myth has also surfaced in historical events. Gold rushes in California (1848–1855), Australia (1851), South Africa (1886), and Alaska's Klondike (1896–1899) were all described by contemporaries as hunts for El Dorado.[93] Accounts of these episodes also regularly referred to gold's pernicious effects.[94] The desire for gold, witnesses observed, "set men's minds on fire," producing a "madness" or "mania."[95] This "gold fever" provoked intense avarice, which, in the words of William E. Connelley, a Kansas pioneer, "changed the American from a conservative, contented citizen satisfied with a reasonable return on his investment to an excitable, restless, insatiable person who wished to realize on the resources of the universe in a day. It was the beginning of our national madness, of our insanity of greed."[96] Men suffering from gold fever felt compelled to pursue their gilded ambitions, regardless of the consequences; the allure of treasure was too powerful to resist. As accounts from the gold fields regularly attested, these ambitions frequently provoked violence. Prospectors turned against each other to grab resources and seize each other's claims.[97]

In the late nineteenth century, the El Dorado myth was extended from gold to oil. The oil fields of western Pennsylvania, the site of the original American oil boom, were nicknamed "Oildorado."[98] So were early California oil fields.[99] The towns of El Dorado, Kansas, and El Dorado, Arkansas, received their names during later petroleum booms. Oil companies also adopted the El Dorado moniker; the Eldorado Drilling Company is located in Oklahoma and the Dorado Oil Company in Texas.[100] Journalists regularly identify oil-producing regions as "El Dorados," particularly after a new strike. Petroleum's nickname—"black gold"—further reinforces the connection. However, other oil-related references to El Dorado are subtler. In the film *Giant* (1956), which takes place in the midst of a Texas oil boom, oilman Jett Rink (James Dean) owns the Conquistador Hotel.[101]

Scholars have also drawn connections between oil and El Dorado. Geographer Michael Watts refers to petroleum producing countries like Nigeria and Ecuador as "petrolic El Dorado[s]."[102] Terry Lynn Karl entitled a chapter of her influential text *The Paradox of Plenty* "Spanish Gold to Black Gold: Commodity Booms Then and Now."[103] Leonardo Maugeri called a chapter of his petroleum history

"The Soviet Implosion and the Troubled Caspian El Dorado."[104] These linguistic flourishes underscore that oil, like gold, is a source of "fantastic wealth."[105] Other authors convey a similar message by applying the term *prize* to international oil competitions. The title of Daniel Yergin's monumental history of the petroleum industry, *The Prize: The Epic Quest for Oil, Money, and Power*, communicates petroleum's value and the extraordinary lengths actors will go to in order to acquire it. Timothy Winegard repeats these tropes in *The First World Oil War* when he refers to certain World War I campaigns as "quest[s] to possess . . . petroleum prizes."[106]

Journalist Ryszard Kapuscinski has offered perhaps the most evocative description of oil's powerful appeal and its pernicious effects. In his words, petroleum "kindles extraordinary emotions and hopes, since oil is above all a great temptation. It is the temptation of ease, wealth, strength, fortune, power. . . . To discover and possess the source of oil is to feel as if, after wandering long underground, you have suddenly stumbled upon royal treasure."[107] The prospect of obtaining this exceptional wealth can drive people to depravity. As Kapuscinski observes, the desire for oil "anesthetizes thought, blurs vision, corrupts."[108] Similarly, during the western Pennsylvania oil boom, commentators observed that prospectors had "oil on the brain" and were regularly struck with "oil fever."[109] This malady, like gold fever, addled their judgment and could have deadly effects.

The film *There Will Be Blood* (2007), adapted from Upton Sinclair's novel *Oil!* (1927), provides the most vivid fictional representation of petroleum greed's destructive power. Set in the oil fields of Southern California in the early 1900s, the film depicts the efforts of prospector Daniel Plainview (Daniel Day Lewis) to amass a fortune from oil.[110] Plainview and other oilmen believe that vast wealth is at stake; in Sinclair's novel, one of the prospectors predicts that petroleum exploration will "yield him a treasure that would make all the oldtime fairy tales and Arabian Nights adventures seem childish things."[111] The oilmen also recognize that petroleum acquisition can incite conflict. One prospector describes a local tussle over the terms of an oil lease in Prospect Hill: "You remember how we heard the racket. . . . Son, that was a little oil war!"[112]

In *There Will Be Blood*, Plainview's oil greed propels him to violence. In the process of amassing his fortune, the oilman swindles and beats his neighbors, blasphemes the church, and murders a business associate. Eventually, Plainview's lust for black gold drives him mad. Ensconced in an enormous mansion with a bowling alley in the basement, the oilman confronts his longtime adversary, preacher Eli Sunday (Paul Dano). After ridiculing Sunday, Plainview murders him by smashing in his head with a bowling pin.[113] Oil inspires the same ruthless, violent behaviors as its gilded counterpart.

The El Dorado myth proposes that actors fight for more than survival. Rather than simply acquiring sufficient resources to fulfill their basic needs, they aspire to fantastic wealth and are willing to go to extraordinary, often violent lengths in order to obtain it. This myth is intuitively plausible, especially for people who possess a more cynical view of human nature. Moreover, to accept the El Dorado myth, it is not necessary to believe that every person or country is greedy. As long as some actors covet treasure, El Dorado provides a compelling explanation for resource-related conflict.

Individually, the Mad Max and El Dorado myths present two distinct pathways from valuable resources to violence: one emphasizing resource need and the other highlighting resource greed. Each therefore incorporates a core motive for conflict that is widely recognized by international relations and conflict theorists. Each myth has also been reiterated for centuries, if not millennia, in popular and academic discourses. These myths have become the conventional wisdom explaining why states fight for valuable natural resources. They therefore provide a durable foundation for the oil wars myth by rendering petroleum-related contention doubly plausible. Since we believe in Mad Max and El Dorado, we believe in classic oil wars.

3

WHY CLASSIC OIL WARS DO NOT PAY

The value of oil has a way of warping human logic.

—Brian Black, *Petrolia* (2000)

The oil wars myth, like the Mad Max and El Dorado myths, rests on a fundamental assumption: that fighting for resources pays. Aggression can fulfill a country's resource needs, as depicted by the Mad Max myth. Or it can satisfy a country's resource greed, as portrayed in the El Dorado myth. Both of these storylines intimate that the payoffs from classic oil wars are high. This assumption makes intuitive sense if we focus on petroleum's value. Of all natural resources, it is the most important for state survival: vital to national military performance and economic productivity. Oil can also be a source of extraordinary wealth; many countries obtain enormous profits by exploiting their petroleum resources. Accordingly, it seems logical to presume that fighting for oil is worth the effort. Given the resource's exceptional value, how could it not be?

Contesting this foundational assumption, I argue that classic oil wars do not pay. Although petroleum is an extremely valuable resource, there are numerous, underappreciated obstacles to seizing and exploiting foreign oil. I group these impediments into four categories: *invasion obstacles*, *occupation obstacles*, *international obstacles*, and *investment obstacles*. Each set of obstacles limits the oil resources and revenue (the *petroleum payoffs*) that an aggressor can obtain by seizing oil-endowed territories. These obstacles constrict petroleum payoffs in a variety of ways, including by decreasing oil production, interrupting oil transportation, depressing oil sales, and raising the costs of oil exploitation. When all four sets of obstacles are taken into account, the petroleum payoffs of international aggression decline dramatically and, often, disappear entirely.[1]

Recognizing these limited payoffs, states are reluctant to fight for oil. Classic oil wars—severe militarized interstate conflicts driven largely by participants' desire to obtain petroleum resources—are therefore implausible. States may engage in mild sparring to advance their petroleum ambitions. Alternatively, oil aspirations may be a marginal, additional incentive in wars fought predominantly for other reasons. However, states' overall willingness to trade blood for oil is far more circumscribed than most people have imagined.

In discussing the obstacles to classic oil wars, I draw extensively from liberal contributions to the value of conquest debate, demonstrating that many of these arguments, previously applied primarily to advanced, industrialized countries, are equally relevant for oil-endowed states in the contemporary international system. I also present historical examples of invasion, occupation, international, and investment obstacles from interstate and intrastate conflicts to illustrate how each set of impediments limits the payoffs of fighting for oil.

Together, the chapter's deductive logic and empirical examples demonstrate that oil war believers, overawed by petroleum's extraordinary value, have overestimated the utility of fighting for it. Classic oil wars are not an effective way to obtain additional petroleum resources or revenue, either for *producer* states—those, like Iraq and Russia, that are net exporters of oil—or for *consumer* states, like France and Japan, that are net importers. Neither type of country can satisfy its oil needs or oil greed by launching a classic oil war. Contrary to popular assumptions, these conflicts do not pay, economically or militarily.

Obstacles to Classic Oil Wars

To align my argument with the value of conquest debate, this chapter initially considers a conquest scenario. In it, an *aggressor* state invades a *target* state to seize its oil resources. The aggressor therefore violates the target's territorial sovereignty to gain control over petroleum reservoirs. If the invasion is successful, the aggressor occupies the seized territory and attempts to exploit its oil or natural gas resources. Since conquest scenarios entail the most severe obstacles to classic oil wars, they provide the strongest challenge to the assumption that these conflicts pay. Later in the chapter, I broaden the discussion to consider a dispute scenario, in which two or more countries fight over petroleum resources in contested territories. The obstacles to classic oil wars decline in this scenario, because resource sovereignty is initially ambiguous. However, the impediments are still much greater than oil war believers have acknowledged.

I also recognize that the severity of each of the four types of obstacles fluctuates, temporally and geographically, in both scenarios. Some time periods have provided more permissive conditions for classic oil wars, with aggression offering more enticing petroleum payoffs. In addition, some prospective aggressors face fewer obstacles to seizing and exploiting foreign oil, making these impediments less of a deterrent. Nonetheless, these variations fail to alter the chapter's overall conclusion: that classic oil wars are not an appealing enterprise.

Invasion Obstacles

Envisioning the oil industry, we immediately think of infrastructure. A timbered oil derrick pierced by a geyser of black crude. A pumpjack nodding up and down, drawing oil out of a subterranean reservoir. An oil refinery's massive, steel tangle of pipes and towers, along with its array of giant, gleaming storage tanks. Metal pipelines snaking through the wilderness. Together, these individual facilities create massive networks of petroleum infrastructure, which must remain operational to sustain a country's oil output.

In a classic oil war, petroleum infrastructure is vulnerable to military attacks. When an aggressor invades oil-endowed territories, it is likely to damage at least a portion of its target's petroleum installations. Most oil facilities are fixed assets, so they cannot flee in the face of an invasion. Nor can they be easily disguised; their function is obvious, even to casual observers. Some facilities, like storage tanks, are also highly flammable. Moreover, oil infrastructure is widely dispersed; oil fields may be located hundreds of miles from processing centers and export facilities. Thus, regardless of which parts of a territory an aggressor attacks, some of its target's oil industry is likely to be in the line of fire. In addition, if the aggressor is an oil producer, its own petroleum infrastructure may be damaged during a classic oil war.

Invasion obstacles are therefore the damage to petroleum infrastructure caused by oil conquest. These obstacles reduce the petroleum payoffs of classic oil wars by diminishing the belligerents' petroleum output. When oil infrastructure is damaged, less crude is extracted from oil fields or processed in refineries. In addition, damage to pipelines, railways, and ports inhibits oil transportation. Because of this destruction, fewer resources are available for domestic consumption and foreign export. Consequently, even if a conqueror successfully seizes foreign oil fields, the resources and revenue that it can immediately capture are limited. Contrary to the expectations of oil war believers, there is no "one-time opportunity for looting" in classic oil wars.[2] An aggressor cannot simply invade an oil-endowed territory, quickly grab its crude, and go home.

This observation flies in the face of previous claims about the value of seizing foreign petroleum resources. Oil war believers have assumed that oil, like other primary commodities, is relatively easy to exploit because, unlike human resources and capital, it cannot flee following a foreign invasion.[3] As Martin McGuire asserts, "Seizing gold, rubber, minerals, lumber, or other natural resources must surely be less costly than enlisting sufficient cooperation from an enslaved population to produce electronic parts, computer programs, or reliable transportation."[4] The immediate gains from conquest, they surmise, must therefore be large. However, this assumption ignores the petroleum industry's complexity and vulnerability. Because of its particular physical and political economic properties, oil is not a highly lootable resource.[5]

There are five ways that invasion can damage oil industry infrastructure, thereby reducing belligerents' oil output and a classic oil war's petroleum payoffs. An aggressor can accidentally or deliberately damage its target's oil infrastructure. A target can accidentally or deliberately damage its attacker's oil infrastructure. Or a target can deliberately damage its own oil infrastructure.

In the first pathway, an aggressor aims to capture the target's oil industry intact, in order to maximize the immediate petroleum payoffs of invading. However, during its seizure of oil-endowed territories, the aggressor may accidentally damage its target's oil installations. Military attacks are unpredictable; bombing and artillery barrages can miss their marks. During World War II, some US bombing campaigns hit their intended objectives only 13 percent of the time.[6] Although targeting technologies have improved dramatically since the 1940s, oil facilities can still be vulnerable if they are located near intended targets. Alternatively, outdated maps may cause an aggressor to inadvertently strike oil installations when it meant to hit something else. Attacks on multiuse facilities, such as roads, railways, and ports, can also reduce petroleum payoffs. Although an aggressor may target those facilities primarily to impede the movements of materiel and military personnel, damaging them will also disrupt oil transportation, causing the aggressor's petroleum payoffs to decline.

In the second pathway, an aggressor deliberately damages its target's oil infrastructure. This pathway initially appears counterintuitive. If a state is prosecuting a classic oil war, aimed at grabbing petroleum resources, why would it harm its target's oil installations? However, for an oil grab to succeed, an aggressor must achieve a military victory. Damaging the target's oil industry can expedite this goal. By interrupting oil extraction, processing, and transportation, attacks on oil infrastructure reduce a target's petroleum output and sales, thereby diminishing its resource revenue and the volume of petroleum products that are available for local military and civilian consumption. This resource denial compromises the target's ability to defend itself, increasing the likelihood of its defeat.

Iraq employed this second tactic in the Iran–Iraq War (1980–1988). In the early days of the conflict, the Iraqi air force attacked Iran's massive oil refinery at Abadan, destroying much of the facility. The Iraqis also hit Iranian installations at Bandar-e Khomeini and on Kharg Island, suspending exports from the latter for two months.[7] As a result of these attacks, Iran's oil output dropped by two-thirds between August and October 1980 and the state's oil revenue declined significantly.[8] These losses may have increased Iraq's chances of victory. However, they also meant that, had the Iraqis seized Iranian oil fields in the war's opening months, their immediate petroleum payoffs would have been small.

The third and fourth pathways entail damage to the aggressor's oil industry. If the conqueror is an oil producer and its facilities are located within range of the target state's artillery or air force, then its infrastructure, too, may be damaged during a classic oil war. In the third pathway, this effect is accidental. If the target retaliates indiscriminately for foreign aggression, striking anything that is within range, some of its random barrages may hit the aggressor's oil installations, thereby reducing the invader's oil output, consumption, and sales. In the fourth pathway, these retaliatory strikes are deliberate. The target aims to damage its attacker's oil facilities in order to lessen the aggressor's chances of victory.

Iran employed the fourth tactic in the Iran–Iraq War. Retaliating against Iraqi attacks, the Iranian air force assaulted pipelines, pumping stations, and refineries at Basra, Kirkuk, and Mosul, as well as Iraq's oil export terminals in the Persian Gulf. These assaults damaged 30 percent of oil infrastructure in Iraq's northern and southern fields, halted oil exports through the Gulf, and suspended pipeline-based exports through Turkey and Syria.[9] Because of these attacks, between August and October 1980, Iraq's oil output dropped by 95 percent.[10] The disruption was so extensive that an Iraqi victory, in the early months of the war, would have been "reverse cumulative"; the state's total petroleum payoffs, from domestic and Iranian fields, would have been lower than its domestic output before the war.[11] Fighting for oil would initially have resulted in a net loss.

The final, and perhaps most effective, way that petroleum infrastructure can be damaged during an invasion is by the target state deliberately attacking its own oil industry. A target's goal, in harming its own facilities, is to deny petroleum resources and revenue to its conqueror. Targets resort to self-sabotage when they believe that their defeat is imminent. Historically, this tactic has been a popular response to invasions of oil-rich territories. Romania employed it during World War I. In November 1916, with German forces poised to seize their oil fields, destruction teams blocked oil wells, wrecked equipment, exploded storage tanks, flooded refineries with petroleum products, and set them ablaze. During World War II, Russian forces destroyed oil fields and infrastructure at Maikop, in the Caucasus, before German troops' arrival. In advance of Japan's invasion of Borneo in 1942, employees of Royal Dutch Shell blew up oil wells, tanks, and docks and demolished the Balikpapan refinery.[12] In southern Sumatra, Standard–Vacuum Oil Company employees attacked the Soengi Gerong refinery, plugged oil wells, and destroyed pipelines.[13] More recently, during Operation Desert Storm (1991), retreating Iraqi forces set fire to over seven hundred Kuwaiti oil wells and damaged drilling machinery, oil gathering centers, and refineries.[14]

Self-sabotage is easy to implement. At Balikpapan, 120 men perpetrated all of the damage over the course of a few days. The destruction in Romania in 1916 was similarly rapid. Retreating Iraqi forces damaged Kuwaiti fields and facilities in spite of the US-led coalition's withering air campaign. In addition, self-sabotage can be quite effective at reducing short-term oil output. During World War I, it took German forces five months to begin extracting oil from damaged Romanian fields. For another six months, output lagged at one-third of its prewar level. In World War II, it took Japan a year to return the Dutch East Indies' oil production to 60 percent of preinvasion levels, in spite of the deployment of over 70 percent of its trained oil field workers to the region. The last oil well fire in Kuwait was extinguished nine months after Iraq's withdrawal, and it took more than two years to fully restore the state's petroleum production.[15]

Invasion obstacles are therefore incurred in a variety of ways. Some of the pathways have greater impact than others in terms of reducing the short-term petroleum payoffs of classic oil wars. Deliberate attacks, for example, are likely to reduce oil output and revenue more than accidental strikes. Moreover, some facilities are easier to damage than others, and some aggressors have greater control over their targeting.[16] Nonetheless, the historical examples suggest that there is significant scope for reducing the immediate payoffs of fighting for oil, especially when multiple pathways operate within the same war.

Invasion obstacles, on their own, can eventually be overcome. Should no other impediments arise, aggressors can repair wartime damage to oil infrastructure, thereby restoring seized territories' petroleum output to preconflict levels. As the historical examples indicate, this process may take a while. In addition, the costs of rebuilding impinge on an aggressor's resource revenue, as the state must recoup reconstruction expenses before it can begin to profit from the restored oil industry. Still, invasion obstacles' transitory nature distinguishes them from the other three types of impediments. Occupation, international, and investment obstacles persist over the long term.

Occupation Obstacles

During the night of June 19, 2008, a small group of armed Nigerian militants raced across the waters of the Gulf of Guinea. Their speedboats were heading toward the Shell Oil Company's Bonga oil platform, seventy-five miles offshore. When they arrived at the facility, the group was unable to access the main control room as planned. However, as a result of the attack, Shell declared force majeure, shutting in 225,000 barrels per day of oil production. With one strike, the Movement for the Emancipation of the Niger Delta (MEND) had taken 10 percent of Nigeria's oil output offline.[17]

The Bonga incident did not occur in the context of a classic oil war. MEND was challenging its own central government, not a foreign country. However, the incident is illustrative of the second set of impediments facing prospective prosecutors of classic oil wars: occupation obstacles. Because of the difficulty of looting oil resources in the immediate aftermath of an invasion, an aggressor that wishes to exploit seized oil fields must be prepared to occupy them for many years, if not decades. During its occupation, the aggressor is likely to face intense local resistance. Hostility toward foreign occupations is intense, and the local population of a conquered territory can channel its opposition into attacks on the petroleum industry. Occupation obstacles are therefore the damage to the oil industry caused by local resistance, as well as the costs an occupier incurs to discourage these attacks. This set of obstacles can significantly reduce the petroleum payoffs of classic oil wars.

Contributors to the value of conquest debate have emphasized local opposition when explaining why conquest no longer pays. They observe that, since the early nineteenth century, nationalism has become an increasingly potent force in domestic and international politics. As individuals' attachment to their national identities has strengthened, so has their hostility to foreign rule.[18] As Klaus Knorr states, "The simple fact is that foreign rule by force of arms is no longer tolerable, and is universally regarded as illegitimate."[19] To demonstrate their opposition, local populations challenge foreign occupations, often forcefully. Since nationalism is "a potent unifying force," this resistance can be very effective in reducing the payoffs of conquest.[20]

Authors have not, however, applied this argument to oil-endowed territories. Some merely observe that nationalism intensifies local resistance in advanced, industrialized societies. Others acknowledge that local resistance also reduces the payoffs of seizing primary commodity-producing countries in general.[21] Yet this obstacle is ignored when oil war believers claim that fighting for petroleum pays.[22]

The omission is unjustified. Nationalism is a potent force in oil-producing states, as well as other countries. By the end of World War II, when oil production began to significantly expand worldwide, the principle of national self-determination had been widely embraced. Many major oil-producing countries, including Saudi Arabia, Iran, Mexico, and Venezuela, have been independent for the entire duration of their oil production. Even the smallest Persian Gulf states, including Kuwait and the United Arab Emirates, became independent in the 1960s and 1970s. Once they were free of foreign rule, the populations of these countries were as likely as others to resist foreign occupation.

The oil industry, moreover, has historically been a flashpoint for nationalist resistance. In many oil-producing countries, the industry is associated with a legacy of foreign exploitation. During the early twentieth century, as the global oil industry was developing, it was dominated by companies from the United States and Western Europe. These "majors" established decades-long concessions agreements with local leaders, which gave the companies control over oil resources, exploration, production, pricing, and sales.[23] Host governments received only a fraction of resource revenue. Over time, resentment of this foreign domination grew and governments began taking action to overturn it. Bolivia nationalized its oil industry in 1937, reclaiming authority over oil resources and production. Mexico followed suit in 1938. In the late 1940s, other producers began to demand a larger share of resource revenue from foreign oil companies. Nationalizations accelerated in the 1960s, and by the mid-1970s, the transition had become a rout; host governments had regained control over their oil.[24]

Having recovered this authority, producers resist giving it up. Governments and local populations of oil-producing states are very sensitive to any developments that undermine national control over oil resources or production decisions. In 2012, when the Mexican government lifted its seventy-five-year-old ban on foreign investment in the state's petroleum industry, domestic opposition was intense.[25] In the wake of the 2015 Iranian nuclear deal (the Joint Comprehensive Plan of Action), the country's most conservative leaders resisted outside involvement in the national oil industry, even as other officials insisted that foreign capital was required to revive production.[26] Saudi Arabia's 2016 proposal for an IPO of 5 percent of its national oil company, Saudi Aramco, also triggered significant domestic opposition.[27] States are loath to cede too much control over national petroleum policy to outside parties.

Local actors, fighting against a foreign occupation, are likely to attack oil infrastructure for pragmatic reasons, as well as symbolic ones. As noted in the discussion of invasion obstacles, the oil industry consists of vulnerable fixed assets. These facilities are therefore logical targets for local resistance. In addition, the oil industry is exceptionally valuable. When deciding whether to sustain an occupation or withdraw, a conqueror will give substantial weight to expected petroleum payoffs. If occupation obstacles are large enough, the aggressor may conclude that maintaining its hold on seized territory is no longer worth the effort. This rationale led Osama bin Laden to instruct al-Qaida members to target Iraq's oil industry in the aftermath of the 2003 US invasion. Bin Laden assumed that, if his supporters were able to persistently impede the country's oil production, the United States would withdraw.[28]

Local opposition groups can target the oil industry in a number of ways. First, they can attack oil company personnel. This tactic is regularly used by rebel groups in civil wars and by transnational insurgent groups. According to one data set, between 1980 and 2011, nonstate actors attacked oil company employees over 350 times.[29] Some of these attacks aim to kill employees, thereby terrorizing the company's remaining personnel and driving them out of the country. In 2004, members of al-Qaida in the Arabian Peninsula employed this tactic, assaulting residential compounds in Khobar, Saudi Arabia, and the offices of ABB Lummus, a chemical engineering company, in Yanbu. The attacks killed twenty-nine civilians, and in their aftermath, ABB Lummus pulled all of its foreign employees out of the country.[30]

In other attacks, rebels hold oil industry personnel for ransom. In Colombia, the FARC (Revolutionary Armed Forces of Colombia) and the ELN (National Liberation Movement) began kidnapping and ransoming oil company employees in the 1980s, as the Caño Limon pipeline was being constructed.[31] Opposition groups in the Niger Delta have also used this tactic extensively since the 1990s, leading some oil companies to shut down operations in the region. Despite the enormous value of the Delta's oil resources, they decided that exploiting them was not worth the security risks.[32]

Second, opposition groups can target oil production and transportation infrastructure. Transportation networks, including pipelines and pumping stations, are particularly popular targets because of their accessibility and the ease of attacks.[33] In Colombia, in 2001 alone, the Caño Limon pipeline was bombed over 170 times.[34] In the Niger Delta, insurgents regularly tap oil pipelines to siphon off crude. Pipeline attacks also proliferated in Iraq following the 2003 invasion, shutting down oil transportation along the country's northern export corridor from 2003 to 2007.[35]

In addition, opposition groups have targeted oil-processing facilities. In 2006, al-Qaida insurgents attacked Saudi Arabia's Abqaiq stabilization facility with explosive-laden trucks.[36] From 2014 to 2015, Islamic State militants attempted to seize and hold Iraq's Baiji oil refinery, which handled one-third of the country's crude oil. Although the insurgents were eventually forced out of the refinery, their clashes with Iraqi security forces were equally effective at interrupting oil processing. The refinery shut down in 2014 and was largely destroyed by the time the prolonged battled ended in October 2015.[37]

The impact of infrastructure attacks on an occupied country's oil output varies. Some types of oil facilities are easier to damage than others. One person armed with simple explosives can render a pipeline inoperable, while a larger, more complex facility can be difficult to disable.[38] Yet rebels have managed to shut down a number of facilities that were previously thought to be impregnable to local attacks. MEND's assault on the Bonga oil platform, for example, was a shock to industry observers, who believed that such a strike was beyond the organization's capabilities.

Some countries' oil industries are also more vulnerable than others. If petroleum infrastructure is highly redundant, with multiple facilities performing the same functions, damage to one portion of the network has limited impact on oil output. Most individual attacks on Saudi Arabian oil infrastructure, for example, would not cause much disruption to the system as a whole. In contrast, in less redundant systems, a single attack can create large interruptions. Sudan has the ability to block the entirety of South Sudan's oil exports simply by shutting down the single pipeline that traverses both states.

Some countries' industries are also better protected than others. Al-Qaida's 2006 attack on Abqaiq had limited impact because of Saudi Arabia's robust defensive systems for its oil installations. Although one truck was able to penetrate the facility's outer perimeter, it was stopped at a second gate, far from the main stabilization facilities, where its explosion caused little damage.[39] The Saudis also closely monitor their pipeline network, enabling them to restore oil flows quickly following attacks.[40] At the opposite end of the spectrum is Nigeria, a country with weak oil industry defenses, where MEND's attacks cut national output by up to 28 percent.[41]

Recognizing this variation, an aggressor can limit the impact of local attacks and sustain oil output by reducing an occupied industry's vulnerabilities. The occupier can increase personnel protection, strengthen infrastructure defense systems, and add network redundancies. In Colombia, assaults on the Caño Limon pipeline dipped significantly from 2002 to 2004, after government forces, supported by the United States, increased their presence in the area.[42] The United States also revived oil transportation in northern Iraq by summer 2007 by establishing heavily defended "pipeline exclusion zones."[43] Alternatively, occupiers can discourage local attacks by buying off insurgents. The Nigerian state's 2009 ceasefire agreement with Delta rebels included provisions for political amnesty, payments to opposition members, and contracts to provide security for the facilities they had previously attacked. Following the accord, assaults on the oil industry declined dramatically.[44]

These defensive measures can therefore sustain oil output. However, they are costly. Consequently, although oil may flow, the economic payoffs of occupying petroleum-endowed territories are still compromised. Moreover, these occupation obstacles persist as long as the conqueror holds seized fields, as any lapses will result in renewed violence. In 2016, after a collapse in international oil prices led to a significant drop in state revenue, the Nigerian government decreased funding for its amnesty program by 70 percent, reducing payments to former militants. In response to these cutbacks, a new opposition group, the Niger Delta Avengers, emerged and revived attacks on the petroleum industry, lowering Nigeria's oil output by 750,000 barrels per day. The interruptions were so significant that, later that year, the government restored previous payment levels, in spite of the burden this placed on the national treasury.[45]

As these examples from civil conflicts show, occupation obstacles can be extensive, even when rebels are only challenging their own governments. In foreign occupations, local resistance is likely to be even more pronounced and effective at disrupting oil output. Whereas, in civil wars, only a portion of the population challenges the ruling government, in an occupation, opposition is likely to be virtually universal.[46] Buying off the population is therefore impractical; the costs are too high. Increasing infrastructure protection is also less productive. Because of their broad-based support, opposition groups can launch more frequent, extensive, and effective attacks. As Eugene Gholz and Daryl G. Press observe, "A successful terrorist campaign in Saudi Arabia, involving repeated attacks on the kingdom's terminals, pipeline junctions, and pumping facilities, could . . . keep vast quantities of oil off the market."[47]

In a foreign occupation, opposition groups also have more tools at their disposal for reducing a conqueror's petroleum payoffs. Since hostility is widespread, the opposition is likely to include current oil industry employees, who can interrupt extraction through work slowdowns and stoppages, generating massive drops in oil production. An antigovernment strike in Iran from 1978 to 1979 reduced national oil output by almost 90 percent. In Venezuela, strikes from 2002 to 2003 lowered oil production by almost 80 percent.[48] Oil company employees can also constrain oil output through sabotage. Since they have access to production and transportation facilities and understand their vulnerabilities, they can design and execute more effective attacks. Several participants in al-Qaida's 2004 attacks in Yanbu, Saudi Arabia, were employees of the engineering company they assaulted; their status gave them easy access to the firm's office.[49]

An occupier can attempt to reduce the impacts of industry strikes and sabotage by monitoring local employees and imposing punishments for work interference and attacks.[50] Alternatively, it can replace local employees with its own or third-country nationals.[51] These responses reduce the likelihood of in-house sabotage and strikes. However, like other defensive measures, they impose additional costs on oil production. Monitoring is expensive, punishment breeds resentment, and foreign workers need time to familiarize themselves with local systems in order to operate them efficiently. Meanwhile, former oil industry employees can continue to use their knowledge of the oil industry to design and execute attacks.

Local resistance significantly lowers the petroleum payoffs of classic oil wars. An occupier must either accept major constraints on its oil output or pay generously in hopes of preventing those losses. These occupation obstacles discourage prospective aggressors from seizing foreign oil resources. Historically, even great powers have been deterred by these impediments. In late 1974 to 1975, during the first energy crisis, US officials debated sending troops to Saudi Arabia or Libya to seize oil fields and enhance American energy security.[52] However, military officers balked at the idea, largely because of the prospect of intense local resistance. Although the United States might be able to seize the fields and restore operations, they asserted, the "problems of maintaining intervention" would be extreme.[53] As a result of occupation obstacles, grabbing foreign oil was not worth the effort.

International Obstacles

When Saddam Hussein invaded Kuwait on August 2, 1990, the international community responded immediately. World leaders condemned the attack, and within a week, the UN passed Resolution 661, imposing trade restrictions on the state. After these sanctions failed to compel Iraq to withdraw, in January 1991, a US-led coalition of thirty-four countries initiated an intense aerial bombing campaign against it. In late February, coalition ground troops entered Kuwait and began to drive Iraqi forces back toward their border. By mid-April, the last Iraqi soldiers had left Kuwait.

Chapter 8 will question whether Iraq's conquest of Kuwait was actually a classic oil war. Nevertheless, the response to this invasion illustrates the third set of impediments to foreign oil grabs: international obstacles. These are the punishments that third-party states and international organizations impose on countries that conquer petroleum-endowed territories. International retaliation reduces the payoffs of classic oil wars by restricting an occupied territory's petroleum production and sales. The aggressor therefore extracts few benefits from seized oil resources while it holds them. It may also be compelled to withdraw from the conquered territories, thereby forfeiting all petroleum payoffs. In addition, if the aggressor is an oil producer, its own petroleum output can be compromised by international retaliation, leaving it worse off than before it launched a classic oil war. Moreover, like occupation obstacles, this set of impediments persists as long as an aggressor holds seized territory.

Liberals have highlighted the danger of international retaliation when explaining why conquest no longer pays. They offer both pragmatic and normative explanations for third parties' willingness to punish international aggression. On a pragmatic level, states prefer to prevent any single country from gaining control over too much land and power, as that state could use its increased might to impose its will on other countries. Consequently, when faced with an international aggressor, other states are likely to balance against it, diplomatically, economically, and militarily.[54]

On a normative level, third parties are encouraged to retaliate against foreign aggression by the international "norm against conquest." This norm, which has guided states' behaviors since the end of World War II, if not earlier in the twentieth century, maintains that seizing another country's sovereign territory is not an acceptable international behavior.[55] To defend the norm, third parties cannot allow conquest to go unpunished. Doing so would set a dangerous precedent for other would-be conquerors, suggesting that they can seize foreign territory with impunity.[56]

When asserting that fighting for oil pays, classic oil war believers have given little thought to international retaliation. However, this set of obstacles is just as relevant—if not more so—for oil conquest as it is for other types of aggression. Third-party states and international organizations have strong pragmatic and normative incentives to resist oil grabs. Pragmatically, countries are wary of any consolidation of authority over petroleum resources since, if one state controls a sizable portion of global reserves, it can manipulate oil production and pricing, and potentially use that power to harm other states.[57] Most countries are therefore inclined to oppose classic oil wars, especially if they target richly endowed territories. In the autumn of 1990, President George H. W. Bush played on these practical concerns to build the large international coalition that participated in Operation Desert Storm. In a speech before Congress, he warned that "an Iraq permitted to swallow Kuwait would have the economic and military power, as well as the arrogance, to intimidate and coerce its neighbors: neighbors who control the lion's share of the world's remaining oil reserves. We cannot permit a resource so vital to be dominated by one so ruthless."[58] Other states concurred; they did not want Saddam Hussein to control over 20 percent of global oil reserves, let alone threaten Saudi Arabia's resources.

On a normative level, conquering foreign territories in order to seize their oil resources is regarded as a particularly immoral act. Although the international community is willing to accept some rationales for foreign intervention, such as the protection of innocent civilians or coethnics, oil grabs are viewed as anachronistic acts of naked state greed. When a Russian submarine planted a national flag in the potentially oil-bearing seabed under the North Pole in 2007, the Canadian foreign minister, Peter MacKay, roundly denounced the action. "This isn't the fifteenth century," he admonished. "You can't just go around putting flags in something and saying 'I'm claiming this territory.'"[59] Similarly, when President

Donald Trump asserted that the United States should have "taken the oil" after invading Iraq in 2003, his statement was widely pilloried.[60]

Oil grabs also violate formal international law. The UN Charter of Economic Rights and Duties of States (1974) explicitly asserts that "every State has and shall freely exercise full permanent sovereignty, including possession, use and disposal, over all its . . . natural resources."[61] Other legal statutes, including the Fourth Hague Convention (1907), assert that foreign governments must not exploit occupied territories' natural resources for their own benefit.[62] Recognizing these normative impediments, Kenneth Waltz described the United States' potential invasion of oil-rich Middle Eastern countries in the mid-1970s as "distasteful" and observed that, if the country attempted it, it "might incur such wrath from so many people that long-term losses would be greater than short-term gains."[63] A senior US military officer described the likely consequences of an American resource grab even more succinctly: "It could create the damndest row in years," he claimed.[64]

Third-party states and international institutions can retaliate for classic oil wars diplomatically, economically, and militarily. The goal of each type of international punishment is to compel the aggressor to withdraw from oil-endowed occupied territories. Diplomatically, third parties can engage in verbal condemnation, like MacKay's rebuke of Russia's North Pole flag plant. These rhetorical critiques can be coupled with travel bans for government officials, withdrawal of diplomatic staff, severance of diplomatic relations, and refusal to cooperate on other issues. International institutions can also deny or suspend an aggressor's membership. Such actions clearly demonstrate third parties' disapproval of international aggression. Yet diplomatic punishments, alone, have little impact on the petroleum payoffs of classic oil wars.

Economic retaliation, in contrast, can substantially reduce an aggressor's petroleum payoffs. Trade restrictions, in particular, lessen the benefits of foreign oil grabs. Third-party states, acting independently or under the aegis of international institutions like the UN, can prohibit oil purchases from occupied territories. If these sanctions are successful, the aggressor will possess additional oil resources but will not be able to sell them internationally.[65] Revenue from seized oil can therefore drop precipitously. If sanctions also prohibit oil purchases from the aggressor's home territories, classic oil wars can be a net economic loss, as the aggressor will sell less oil and collect less petroleum revenue than it did before the invasion. Iraq experienced both of these losses in 1990, when UN sanctions caused its oil exports from occupied Kuwait and its home territories to drop to almost zero.[66] These revenue shocks also persisted for years, because the UN and other parties sustained their sanctions long after Iraq's defeat. The state's oil output only began to revive in 1997, with the implementation of the UN's oil-for-food program.[67] Thus, international economic retaliation can generate lengthy, as well as large, reductions in petroleum payoffs.

International military retaliation can also drastically reduce a classic oil war's petroleum payoffs. Military operations may target the aggressor's armed forces to compel them to withdraw from occupied territories, as occurred in Operation Desert Storm. They can also target occupied oil production facilities and transportation networks to diminish an aggressor's petroleum consumption and sales, thereby hastening its defeat and its withdrawal from seized territories. During World War II, Allied bombing campaigns employed the latter tactic, striking German-occupied oil fields in Ploieşti, Romania, to limit the Nazis' petroleum output.[68] The United States also bombed Japanese tankers transporting oil from the occupied Dutch East Indies to Japan's home islands. As a result of these attacks, the Axis powers reaped limited benefits from their seized oil reservoirs.

In addition to eliminating petroleum payoffs from occupied territories, international military retaliation can cause an oil war to be reverse cumulative. If third parties damage the aggressor's domestic oil industry, as well as infrastructure in occupied territories, the state will be worse off following a defeat than it was before its invasion. Again, the coalition response to Iraq's invasion of Kuwait is illustrative. Not only was Iraq forced out of Kuwait; its own oil industry was pummeled by the US-led bombing campaign. By the end of Operation Desert Storm, 90 percent of Iraq's refining capabilities had been taken offline.[69] The country's postwar petroleum output was therefore dramatically lower than it had been before seizing foreign oil. This physical damage would have limited Iraq's resource output after the war, even in the absence of economic sanctions.

Taken together, international economic and military retaliation can severely constrain, or even reverse, the petroleum payoffs of classic oil wars. As a result, even great powers, which appear to be less vulnerable to international punishment, are discouraged from seizing oil-endowed territories. During the mid-1970s debate about seizing Middle Eastern oil fields, a Congressional Research Service study emphasized the risk of retaliation by the Soviet Union. If the United States attempted to seize regional oil resources, the report asserted, military action by Soviet air and ground forces was "a distinct possibility." Soviet attacks were expected to target oil production facilities within occupied territories, as well as tankers shipping oil from occupied territories to the United States and its allies.[70] Recognizing these international obstacles, the Ford administration refrained from an oil grab.

Investment Obstacles

In 1995, thirteen years after the Falklands War (1982), the Falkland Islands government invited bids for oil companies to explore around the contested archipelago. The area was believed to contain petroleum resources but, up to that point, no wells had been drilled or discoveries made. The Falklanders opened bidding for blocks north and south of the islands. The northern blocks, which were located farther from Argentina, attracted bids from fourteen oil companies. In contrast, no companies bid for the southern area.[71] Although the United Kingdom had forcefully demonstrated its willingness to defend the islands only thirteen years earlier, oil companies were still hesitant to invest too close to Argentina, recognizing that their exploration rights might be challenged or overturned.

Their hesitation is indicative of the fourth set of impediments to classic oil wars: investment obstacles. In addition to triggering negative responses from local populations, third-party states, and international institutions, oil wars alienate oil companies. Investment obstacles are the losses generated by foreign companies' reluctance to participate in oil projects in occupied territories. As profit-seeking actors, these companies prefer to avoid unstable investment environments. Yet many would-be conquerors require foreign capital to explore for, develop, extract, and market seized petroleum resources. These aggressors are therefore faced with a conundrum. They must either forgo outside investment, reducing their oil output from occupied territories and potentially their home territories, or accept unfavorable contract terms in order to attract foreign partners. With both choices, the petroleum payoffs of classic oil wars decline.

Previous contributors to the value of conquest debate have recognized that foreign companies avoid investing in occupied territories. Norman Angell observes that invasions create intense insecurity. They put contracts at risk and reduce faith in investments, as no one knows how "alien governors only concerned to exact tribute" will manage conquered territory.[72] Stephen Brooks elaborates on this credible commitment problem, noting that "any extractive conqueror will not be able to assure foreign investors that it will abide by . . . policies and will not seize assets of MNCs [multinational corporations], extract excessive rents from them, or generally shift policies in ways that reduce the cost-effectiveness of investments."[73]

Aggressors appear unreliable because they have violated the norm against conquest. If they are indifferent to international rules that prohibit grabbing foreign territory, they may also ignore private property norms, seize assets, cancel contracts, and implement dramatic, unpredictable changes in fiscal policy.[74] In response to this uncertainty, foreign investment in conquered territories is likely to decline; as Brooks puts it, "There are strong reasons to expect the flow of inward FDI [foreign direct investment] to decline markedly . . . after it is vanquished by an extractive conqueror."[75] Alternatively, the aggressor will be forced to solicit foreign participation on "usurious and extortionate" terms.[76]

Brooks recognizes that investors' fears are likely to be particularly pronounced for FDI in physical infrastructure because of the difficulty of moving these assets.[77]

However, neither he nor other oil war believers consider investment obstacles when they claim that fighting for oil pays. Again, this is an unjustified omission. Foreign investment is critical to many conquerors' prospective petroleum payoffs. Although some countries possess sufficient domestic capital, equipment, and technical expertise to develop seized oil resources on their own, many do not. Even Russia, one of the world's leading petroleum producers, has required foreign oil company participation to execute its most complex oil and gas projects in areas like the Arctic.[78] These countries must attract foreign investment if they want to obtain petroleum payoffs from classic oil wars.

Oil companies, however, are likely to be exceptionally cautious about operating in unstable investment environments, because of the petroleum industry's physical and political economic properties. As observed earlier, the industry is comprised of immobile assets, such as wells, refineries, and export terminals, which makes it unusually physically vulnerable. In addition, oil exploration and development are extremely expensive activities. Operating an offshore oil rig, for example, costs between $250,000 and $500,000 per day. Individual oil projects regularly cost more than $1 billion. Moreover, capital expenditures are highest in the early stages of development, before oil has begun to flow. Consequently, in order to turn a profit, investors must sustain their participation in conventional oil projects for many years, if not decades. If political or economic conditions deteriorate during that time period, the company's profits are jeopardized.

There are numerous ways that conquest can dampen foreign oil companies' enthusiasm for investing in occupied territories.[79] First, there is the risk of recidivism.[80] If an aggressor is overthrown and the target state regains political authority, a new investor can lose everything. The returning government is likely to cancel exploration and production contracts and confiscate oil facilities to punish the company for doing business with the occupying regime. Moreover, since that regime was illegitimate, the investor will garner little international sympathy for its losses, which limits its ability to seek compensation.

Second, even if an aggressor retains power, oil investments in occupied territories remain dubious endeavors. The occupying regime itself poses a significant threat to company profits. It may cancel or arbitrarily renegotiate contracts, forcing oil companies to accept less favorable terms. Alternatively, it may impose prohibitively high taxes, reducing an investor's net revenue. In a worst-case scenario, the occupying regime can nationalize the occupied territory's oil industry, taking control of operations and expropriating foreign investors' physical assets.

Foreign oil companies have limited recourse under these circumstances. They cannot repatriate their facilities or shift extraction to new locations; the oil stays where it is. In addition, multinational companies' efforts to obtain compensation for such losses have historically produced mixed results. When Bolivia, Mexico, and Iran nationalized their oil industries, in 1937, 1938, and 1951, respectively, the foreign oil companies that had been operating in these countries received minimal compensation.[81] More recently, the Spanish oil company Repsol received only half the sum it demanded from Argentina after the country seized its shares in the national oil company, Yacimientos Petrolíferos Fiscales.[82] Likewise, Venezuela refused to implement an arbitral award issued by the World Bank's International Center for Settlement of Investment Disputes in 2014, even after the $1.6 billion settlement with ExxonMobil was reduced to $188 million in a major win for Caracas.[83] Oil companies' ability to obtain compensation will be even more limited in the context of occupation. Given aggressors' indifference to international norms, they are unlikely to respond to lawsuits.

Third, foreign oil companies are reluctant to invest in occupied territories because of occupation obstacles. When local opposition groups attack oil installations and personnel, investor profits, as well as aggressor payoffs, decline.[84] Like conquering states, oil companies must either accept lower petroleum output, as a result of local resistance, or expend money to sustain production: funding ransom payments, paying larger insurance premiums, offering higher salaries to attract employees, and financing greater personnel and facility security. If these costs are prohibitive, foreign oil companies will refrain from investing in new petroleum projects and pull out of existing ones. In Colombia, companies avoided new investments around the Caño Limon pipeline, even after joint US–Colombian military operations increased regional security.[85] In Nigeria, MEND's attacks compelled Shell and Chevron to withdraw from onshore oil projects.[86] After civil conflict reignited in South Sudan in 2014, a number of companies, including Total, ExxonMobil, and the Indian National Oil Company, withdrew from the country.[87] Moreover, these examples are drawn only from civil wars. In the context of foreign occupations, where local opposition is likely to be more intense, investment capital will be even more elusive.

Fourth, oil companies are deterred from investing in occupied territories because of international obstacles. International economic sanctions may directly prohibit foreign participation in oil projects in occupied territories. Alternatively, they may impede resource development indirectly, by barring technology transfers or trade in goods and services. Both of these practices limited investment in the Russian Arctic following the state's 2014 incursions into Ukraine, thereby slowing regional exploration and development.[88] Even if companies are allowed to invest in occupied territories, sanctions campaigns can impinge on their profits by prohibiting oil exports. Unable to market seized oil internationally, companies' profits will decline precipitously. Meanwhile, should international military retaliation successfully overturn the occupation, restoring the target state's political authority, companies may lose everything.

All of these concerns have a chilling effect on foreign investment in petroleum projects in conquered territories. Hoon Lee found that, when countries participate in interstate and intrastate armed conflicts, FDI in their oil industries declines.[89] Projects in occupied territories are likely to have even greater difficulty attracting international financing. As a result, aggressors will not be able to obtain the capital, equipment, and technical expertise they need to exploit seized oil resources. Alternatively, they will have to accept unfavorable contract terms that give the government a smaller share of oil resources and revenue. If an aggressor is highly dependent on outside assistance to exploit seized oil, investment obstacles can cause a severe drop in the petroleum payoffs of international oil grabs.

A Dispute Scenario

Together, invasion, occupation, international, and investment obstacles drastically limit the petroleum payoffs of classic oil wars. As a result, fighting for oil does not pay nearly as much as international relations scholars, popular commentators, and the general public have assumed. If circumstances are extremely unfavorable, petroleum payoffs disappear entirely. Aggression can even leave a conqueror worse off, in terms of oil resources and revenue, than it was before its attack. Given these limited payoffs, the desire to seize oil resources cannot be a strong motivation for international conquest. This limitation renders classic oil wars, in which petroleum ambitions are a significant motivator for aggression, implausible.

There are conditions, however, that reduce the impediments to classic oil wars. The most important of these is the dispute scenario mentioned at the beginning of this chapter. In contrast to a conquest scenario, where an aggressor seizes oil-endowed territory that clearly belongs to another state, in a dispute scenario, two or more countries fight over areas where sovereignty is uncertain. For example, in much of the South China Sea, multiple states can legitimately claim potentially petroleum-endowed territories. In this type of scenario, where states are competing over resources rather than stealing them, the impediments to fighting for oil are less severe. However, petroleum payoffs are also less certain.

In disputed territories, oil resources are often underdeveloped. Interstate competition impedes petroleum exploration and development, as each claimant country is likely to resist other claimants' unilateral projects to exploit contested oil reservoirs. In addition, states that are locked in an acrimonious territorial dispute have difficulty coordinating exploration. Consequently, in dispute scenarios, countries tend to compete over prospective oil resources rather than known ones. This underdevelopment reduces invasion obstacles to classic oil wars, because there is little petroleum infrastructure for belligerents to destroy. Similarly, the lack of oil infrastructure initially reduces occupation obstacles, because local opposition groups have no petroleum industry targets. In addition, the populations of contested territories may have weaker attachments to a particular government, so their opposition to a new authority can be relatively muted. Locals may even have a stronger affinity for the new regime than the old one, further curbing their resistance. If contested territories are uninhabited, as is the case for some small islands and maritime areas, occupation obstacles are completely absent; there is no one to challenge a new political authority.

These reductions in invasion and occupation obstacles are counterbalanced, however, by the increased uncertainty of petroleum payoffs in dispute scenarios. Since oil resources tend to be prospective rather than known, aggressors do not know precisely what they are fighting for. In the short term, they will receive no petroleum payoffs from their attacks. In the long run, the territory they seize may reveal a petroleum bonanza. Or it can produce nothing but dry holes. Claimants will therefore hesitate to attack even if invasion and occupation obstacles are low.

International obstacles also continue to deter aggression in dispute scenarios. The international community is still likely to censure the forceful seizure of territory even if areas are contested. Although this aggression does not violate the norm against conquest, it flouts other international principles: in particular, prohibitions against the use of force to resolve interstate disputes.[90] Third-party states also retain pragmatic incentives to prevent the consolidation of control over global oil resources, whether they are prospective or known. Thus, international punishments for seizing contested, potentially oil-endowed territories are still substantial.

Investment obstacles also fail to decline significantly in dispute scenarios. Although a reduction in occupation obstacles may encourage foreign oil companies' participation in petroleum projects, these actors still face the threat of future reversals of political authority, as many territories remain contested even after one country forcefully asserts its sovereign control. In addition, investors must still deal with a norm-violating regime, which may be inclined to forcefully renegotiate contracts or seize private assets. Moreover, when oil resources are underdeveloped, investors' costs are higher and their payoffs less certain. Companies must accept the expenses and risks of oil exploration rather than simply exploiting discovered fields. All of these concerns sustain investment obstacles in dispute scenarios.

The collective obstacles to classic oil wars therefore decline mildly in dispute scenarios, when compared with conquest scenarios. However, the payoffs from aggression decline as well. Consequently, states' willingness to fight for oil is likely to be low in dispute, as well as conquest, scenarios. Given these impediments, states are unlikely to initiate classic oil wars; these conflicts are simply not worth the effort under any conditions. However, it is possible that oil aspirations will inspire minor interstate confrontations, particularly in dispute scenarios, as states may conclude that mild conflicts, launched in favorable conditions, pay. Alternatively, states may fight in oil-endowed territories, but for other reasons.

4
SEARCHING FOR CLASSIC OIL WARS

Do countries fight for oil resources? Although the previous chapter argued that classic oil wars do not pay, that does not mean that they never occur in practice. Leaders, under the sway of the oil wars myth, may miscalculate the impediments to seizing oil, overestimating petroleum payoffs and launching inefficient attacks. Alternatively, authorities may ignore the costs and benefits of conflict altogether, eschewing rationalist approaches to foreign policy decision making.[1] It is therefore imperative to empirically assess whether classic oil wars actually occur. To be compelling, such an evaluation must also overcome the two major deficiencies of existing classic oil war research: case studies' lack of generalizability and statistical analyses' neglect of causal relationships.

To overcome the first shortcoming, this chapter presents an appraisal of over six hundred militarized interstate disputes (MIDs) from 1912 to 2010: almost a century of potentially oil-driven international conflict. This is not a universal sample of possible classic oil wars. However, it far exceeds the number of cases examined in previous qualitative research, heightening confidence in the study's findings. To surmount the second shortcoming, the analysis investigated states' motives for conflict: specifically, whether the desire to obtain control over oil or natural gas resources inspired their international aggression. To make this determination, I identified each MID's geographic location and ascertained whether contested areas contained known or prospective oil or gas reservoirs. For the 180 conflicts involving hydrocarbon-endowed territories, I evaluated how petroleum ambitions contributed to leaders' decisions to engage in international aggression.

My analysis found that states' willingness to fight for oil resources is, in fact, highly circumscribed. With one possible exception, I identified no classic oil wars—that is, severe militarized interstate conflicts driven largely by participants' desire to obtain petroleum resources. Instead, as chapter 3 predicted, states either engaged in mild, oil-oriented confrontations or they fought for other reasons. Evidently, when confronted by the actual prospect of severe interstate violence, leaders are capable of overcoming the intellectual inertia generated by the oil wars myth to recognize that classic oil wars do not pay.

Rather than prosecuting classic oil wars, states engage in four types of militarized conflicts in hydrocarbon-endowed territories: *oil spats*, *red herrings*, *oil campaigns*, and *oil gambits*. Oil spats are minor, usually nonfatal military confrontations inspired by petroleum ambitions. Put simply, they possess the "oil" element of classic oil wars, but not the "war" element. The other three categories fail to satisfy the "oil" component; the desire to obtain petroleum is not a significant cause of conflict. Red herrings are motivated predominantly by other issues, including domestic politics, shifting regional power balances, hegemonic aspirations, national pride, the desire for political independence, and contested territories' other economic, strategic, or symbolic assets, rather than oil. Oil campaigns, in contrast, target petroleum resources. They are also, unlike oil spats, very intense, resulting in hundreds, if not thousands, of fatalities. However, these severe interstate conflicts occur in the midst of ongoing wars that were not themselves caused by petroleum ambitions.

Oil can therefore influence the trajectories of wars that are already under way. However, it does not start them. The one possible exception is the study's unique oil gambit: Iraq's invasion of Kuwait. This was a severe conflict, launched in peacetime, that targeted petroleum resources. Nonetheless, the label "oil gambit" captures the desperate, instrumental character of Iraq's invasion more accurately than the label "classic oil war." Alternatively, should we choose to call the invasion a classic oil war, we must also acknowledge that these conflicts look very different from what most of us have imagined—and are initiated far more reluctantly.

Method

Searching for classic oil wars entails multiple methodological challenges. First, if the argument presented in chapter 3 is correct—if states avoid fighting to obtain petroleum resources—then classic oil wars are dogs that don't bark. Consequently, they will be largely absent from the historical record, their only potential traces being attacks that were considered but rejected. Identifying enough of these forgone acts of aggression to ground a rigorous, generalizable analysis is a dubious task. To overcome this obstacle, I examined actual cases of militarized conflict and searched for petroleum motives. If oil ambitions made little contribution to conflict initiation, my argument was supported. If, to the contrary, oil aspirations inspired many acts of aggression, especially major ones, my argument would be undermined. Adopting this approach, of examining historical conflicts rather than forgone aggression, also enabled my study to respond to an additional question: If states do not fight classic oil wars, what do they do *instead*?

To assemble a relevant sample of cases for answering this question, I employed the Correlates of War Project's MID data set, the most comprehensive catalog of twentieth-century international conflicts.[2] The data set enabled me to conduct a thorough evaluation of oil's contributions to interstate violence, because MIDs vary widely in intensity. They include interstate wars, conventionally defined as conflicts between two or more independent states that result in over one thousand battle deaths. However, the data set also catalogs less severe militarized episodes, including threats to use force, shows of force, alerts, mobilizations, blockades, occupations, attacks, and clashes.[3] These other forms of interstate conflict all result in fewer than one thousand battle deaths, and the majority of them are nonfatal. To identify conflicts as classic oil wars, I set a minimum fatality threshold of twenty-five battle deaths over the course of the entire MID. This threshold is far lower than the conventional measure for interstate wars and, consequently, more inclusive. Yet it still conforms to previous classic oil war definitions, which characterize these conflicts as "particularly intense."[4]

The MID data set also facilitates an examination of oil's contributions to international conflict by identifying the primary and secondary issues that states were fighting for in each episode: *territory*, *regime or government*, *policy*, or *other*. To evaluate oil's impact, I examined all MIDs in which *territory* was the primary or secondary motive of at least one dispute participant. I focused on these cases because oil-related contention is inherently territorial. To exploit petroleum resources, a country must control the areas where they are located. In addition, I examined all of the MIDs with *other* motives to determine whether they were, in fact, provoked by states' petroleum aspirations.[5]

My study covers the 1912–2010 time period. Commercial oil production began a half century earlier, in the oil fields of northwestern Pennsylvania. However, it was not until 1912 that Winston Churchill, then First Lord of the Admiralty, decided to switch the British navy's primary fuel from coal to oil.[6] This transition transformed petroleum into a vital strategic resource, significantly increasing its value for consumer and producer states. The study's start date reflects this shift in oil's status. The end date was determined by MID data availability.

Between 1912 and 2010, states prosecuted 617 *territorial* and 26 *other* MIDs. To determine whether oil ambitions inspired each conflict, it was necessary, first, to identify the cases in which states were competing over territories that contained hydrocarbon resources. To determine which geographic areas states were fighting for in each episode, I consulted the MID Locations data set and considered aggressors' territorial claims at each conflict's outset.[7] To identify areas containing hydrocarbon resources, I began with the Petroleum Dispute Dataset, the leading public data source on *known* oil and natural gas fields.[8] However, this data set provides little information about territories' *prospective* hydrocarbon endowments—that is, oil or gas resources that are believed to exist but still unconfirmed.[9] Previous studies of classic oil wars have largely overlooked prospective petroleum resources. However, from a theoretical perspective, there is no reason to exclude them; states can fight over rumored reservoirs, as well as known ones. I therefore supplemented the Petroleum Dispute Dataset with a wide variety of sources, including journalistic accounts, historical case studies, and reference guides on armed conflict, territorial disputes, and international boundaries. I classified contested areas as *hydrocarbon-endowed* if they contained known oil or natural gas resources or belief in these resources' presence was widespread.[10]

In total, my preliminary analysis identified 180 MIDs involving hydrocarbon-endowed territories. This is a significant number, which explains why statistical studies of oil–conflict relationships sometimes produce positive results; a purely correlational analysis would identify most of the cases as oil-related. Table 4.1 catalogs these 180 MIDs, grouping them based on their participants and the hydrocarbon-endowed territories they were contesting. As the table indicates, many states have perpetrated multiple MIDs over hydrocarbon-rich areas. Some dyads prosecuted repeated MIDs in the same hydrocarbon-endowed territory. Others prosecuted MIDs over multiple, distinct hydrocarbon-endowed areas.[11] China and Vietnam, for example, engaged in MIDs in the Gulf of Tonkin, around the Paracel Islands, and near Vanguard Bank.

Most MIDs involving hydrocarbon-endowed territories are bilateral. However, in some contested areas—specifically, the East China Sea and South China Sea—MIDs have occurred between three or more states.[12] Table 4.1 groups these cases together to reflect their interconnectedness. For the same reason, the table clusters the MIDs associated with World War I and World War II.[13] It also lists some conflicts in bold to indicate that they resulted in more than twenty-five battle deaths. These thirty-nine severe conflicts are particularly significant for this study, as they are the potential classic oil wars.[14] No other conflicts were deadly enough to qualify for this designation, even with the study's low fatality threshold. Nonetheless, I investigated decision makers' motives for aggression in all 180 MIDs to obtain a broader understanding of how oil resources affect interstate conflict.

TABLE 4.1 Militarized interstate disputes (MIDs) involving hydrocarbon-endowed territories

PARTICIPANTS	CONTESTED HYDROCARBON-ENDOWED TERRITORIES	MID DATES (ID NUMBER)
Albania–Yugoslavia	Albania	**1949–51 (2328)**
Argentina–Chile	Antarctica; Strait of Magellan	1977–78 (2081); 1978–79 (2082); 1980–81 (2083)
Argentina–United Kingdom	Falkland/Malvinas Islands	1976 (363); **1982 (3630)**; 1983 (3064)
Argentina–Uruguay	Rio de la Plata estuary	1969 (1172); 1973 (1808)
Azerbaijan–Iran	Caspian Sea	2001 (4317)
Bahrain–Qatar	Hawar Islands	1986 (2572)
Belize–Guatemala	Belize	1972 (2319)[a]; 1975 (360); 1977 (2139); 2000 (4151)
Bolivia–Paraguay	Chaco Boreal	**1931–35 (1027)**; 1935 (1211); 1936 (1028); 1936–37 (2134); 1937 (2135); 1937 (1082); 1938 (1029)
Bosnia and Herzegovina–Croatia–Yugoslavia	Bosnia and Herzegovina; Croatia	**1992 (3555)**; **1992–93 (3557)**; **1992 (3556)**; 1993 (4340); **1993–94 (4341)**
Burkina Faso–Mali	Agacher Strip	1974–75 (1411); **1985–86 (2583)**; 1986 (3629)
Cameroon–Nigeria	Bakassi Peninsula	1994 (4119); **1995 (4165)**; **1996–97 (4166)**; 1998 (4250); 2005 (4380)
Chad–Libya	Aouzou Strip; Chad	1976–77 (1337); **1977–80 (3631)**; 1994 (4164)
Chad–Nigeria	Lake Chad	1995 (4068)
China–Indonesia	Natuna Islands	1996 (4063)
China–Japan	Manchuria	**1931–33 (129)**; **1937–41 (157)**
China–Japan–Taiwan	East China Sea, including Senkaku/Diaoyu Islands	1978 (2966); 1995 (4025); 1995 (4061); 1996 (4062); 1996 (4026); 1999 (4180); 2005 (4470); 2010 (4491)[b]
China–Malaysia–Philippines–Taiwan–Vietnam	Spratly Islands	1975 (1728); 1980 (3117); 1983–84 (3616); 1987 (2780); **1988 (2749)**; 1994 (4024); 1994 (4331); 1995 (4027); 1995 (4060); 1996–97 (4028); 1998 (4329); 1998–2000 (4128); 1998 (4328); 1999 (4330); 2001 (4279)[c]
China–Taiwan	China; Taiwan	**1949–50 (633)**; 1951–52 (2052); **1953–56 (50)**; **1958 (173)**
China–Vietnam	Paracel Islands; Gulf of Tonkin; Vanguard Bank	**1974 (355)**; 1993 (4029); 1994 (4030)
Colombia–Nicaragua	San Andrés y Providencia	1980 (3120); 1994 (4145); 2001 (4263)

(*continued*)

TABLE 4.1 (continued)

PARTICIPANTS	CONTESTED HYDROCARBON-ENDOWED TERRITORIES	MID DATES (ID NUMBER)
Colombia–Venezuela	Guajira Peninsula; Gulf of Venezuela; Los Monjes Islands	1982 (2323); 1986 (2356); 1987 (2812)
Democratic Republic of the Congo–Uganda	Lake Albert	2007 (4392)
Egypt–Israel	Sinai Peninsula	**1966–67 (1035)**; **1967–70 (1480)**; 1970–71 (3387); **1971–73 (1046)**; 1973–75 (3380)
Egypt–Sudan	Hala'ib Triangle	1995 (4134); 1996 (4288)
Equatorial Guinea–Gabon	Corisco Bay Islands	1972 (1340)
Eritrea–Yemen	Hanish Islands	1995–96 (4121); 1997 (4132)
Estonia–Soviet Union	Estonia	1918–19 (2605)
Ethiopia–Somalia	Ogaden region	1923 (1669); 1930 (406); 1931 (407); **1934–36 (111)**[d]; 1960 (1423); 1961 (1421); **1963–64 (1425)**; 1965 (2066); 1966 (2067); 1973 (2068); 1974 (1427); 1975–76 (1428); **1977–78 (2069)**; **1978–79 (2070)**; **1980–81 (2071)**; **1982–83 (2072)**; 1984 (2073); **1984–85 (2074)**; **1985 (2075)**
Greece–Turkey	Aegean Sea	1974 (1292); 1974 (2173); 1976 (1289); 1978 (2174); 1981–82 (2175); 1982–84 (2176); 1984–85 (2177); 1986–87 (2179); 1989 (3909); 1995–96 (4092); 1997 (4323); 1997–98 (4193); 1999 (4133); 2000–2003 (4320); 2004–2005 (4423); 2006 (4431)
Guyana–Suriname	Maritime boundary	1977–78 (2326)
Guyana–Venezuela	Essequibo Province	1981–82 (2237); 1982 (3085); 1999 (4260)
Honduras–Nicaragua	Gracias à Dios Province	1957 (1173); 1995 (4012)
Indonesia–Malaysia	Brunei; Sarawak	**1963–65 (1070)**
Indonesia–Netherlands	Western New Guinea	1951–52 (1023); **1953 (2000)**; 1957 (2019); 1957–59 (1024); **1960–62 (1021)**; 1960 (1604); 1961–62 (1022)
Indonesia–Portugal	East Timor	1975–76 (1450)
Indonesia–Vietnam	Natuna Islands	1980 (3610)
Iran–Iraq	Border zone; Khuzestan Province	1934–35 (2103); 1971 (2110); 1971–72 (1135); **1979–80 (2114)**; **1980–88 (2115)**; 2009–10 (4546)

TABLE 4.1 (continued)

PARTICIPANTS	CONTESTED HYDROCARBON-ENDOWED TERRITORIES	MID DATES (ID NUMBER)
Iran–United Arab Emirates	Abu Musa and Tunb islands	1992 (3567)
Iraq–Kuwait	Kuwait	1961–62 (122); 1967 (3172); 1972–73 (1612); 1975 (1613); **1990–91 (3957)**; 1992–94 (3568); 1994–95 (4269); 1996–97 (4272); 1999 (4274); 2000 (4275)
Japan–South Korea	Takeshima/Dokdo Islands	1996 (4126); 2005 (4468)
Kyrgyzstan–Uzbekistan	Fergana Valley	1999 (4177)
Libya–Tunisia	Gulf of Gabès	1977 (3014)
Morocco–Spain	Western Sahara	1957–58 (1117)
Saudi Arabia–Yemen	Land boundary; Red Sea islands	1994–95 (4114); 1997–98 (4203)
Turkey–United Kingdom	Mosul Province	1925–26 (3185)
World War I[e]	Albania; Austria; Czechoslovakia; France; Germany; Italy; Ottoman Empire; Poland; Romania; Russia; Yugoslavia	1914 (394); 1918 (1262)[f]
World War II	Albania; Austria; Burma; Czechoslovakia; Denmark; Dutch East Indies; France; Germany; Hungary; Italy; Manchuria; Netherlands; northern Borneo; Poland; Romania; Soviet Union; Yugoslavia	1938 (11); 1938 (12); 1939 (2302); **1939–45 (258)**; **1939 (108)**; **1939 (169)**; 1939–40 (3701); **1940 (3706)**; **1940–41 (3813)**; 1940–41 (3822); 1940–41 (3825); 1945 (2725)[g]

Note: Boldface indicates that the conflict resulted in more than twenty-five battle deaths.

[a] The participants in MIDs 2319, 360, and 2319 were Guatemala–United Kingdom (Belize's predecessor).

[b] MID participants were China–Japan (2966, 4061, 4062, 4180, 4491) and Japan–Taiwan (4025, 4026, 4470).

[c] MID participants were China–Democratic Republic of Vietnam/Vietnam (2780, 2749, 4328); China–Malaysia–Vietnam (3616); China–Philippines (4027, 4028, 4128, 4279); Malaysia–Philippines (3117); Philippines–Taiwan (4024); Philippines–Vietnam (4329, 4330); and Taiwan–Vietnam (4331, 4060).

[d] The participants in MIDs 1669, 406, 407, and 111 were Ethiopia–Italy.

[e] The MID data set does not code World War I (MID 247) as a *territorial* conflict. I have nonetheless listed the hydrocarbon-endowed states that were contested during the conflict, because some of the war's campaigns targeted oil resources.

[f] The participants in MIDs 394 and 1262 were Albania–Italy.

[g] MID participants were Albania–Italy (108); Austria–Germany (11); Czechoslovakia–Germany (12, 2302); Denmark–Germany (3706); Germany–Netherlands (3701); Germany–Romania (3825); Germany–USSR (3822); Germany–Yugoslavia (3813); Japan–Mongolia (2725); and Poland–Soviet Union (169). MID 258 is the main record for World War II.

A second methodological challenge, when searching for classic oil wars, is the difficulty of assessing decision makers' motives for aggression. Few—if any—conflicts are monocausal; states have multiple reasons for international contention, which are difficult to disentangle and rank. In addition, leaders may not be entirely sure of their motives, records of decision-making processes are incomplete, and many sources are unreliable. Journalists, for example, tend to default to classic oil war interpretations of any confrontations that occur in the vicinity of known or prospective hydrocarbon resources. Moreover, reporters' geographic knowledge is sometimes lacking, so they may misrepresent conflict and resource locations. Periodically, journalists suggest that conflicts are connected to hydrocarbons even though the confrontations and resource deposits they describe are located hundreds of miles apart.[15]

Leaders' public statements can also be inaccurate indicators of hydrocarbons' contribution to conflict initiation. Officials are aware that seizing another state's oil or gas resources is considered an illegitimate reason for foreign aggression. Consequently, they have strong incentives to conceal their hydrocarbon ambitions, perhaps by cloaking them in more acceptable rationales for foreign intervention, such as protecting coethnics or reclaiming previously held territories. Leaders are also likely to accuse their adversaries of perpetrating oil grabs, in order to discredit their motives and rally international support to their own side of a conflict. If multiple MID participants adopt this tactic, disentangling the causes of contention becomes even more challenging.

To overcome these obstacles, I employed methods pioneered by previous researchers, examining the issues that inspire international conflict.[16] I reviewed historical accounts of each MID, using case studies and reference guides to triangulate between previous authors' assessments of hydrocarbons' contribution to each conflict's onset. When secondary sources failed to firmly establish decision makers' motives, I consulted published primary sources and archival materials documenting leaders' private meetings, where they could express their aspirations more freely. I also evaluated whether each MID was initiated in a manner that was consistent with an oil motive. Did early attacks target hydrocarbon-endowed territories? Was the aggressor determined to hold those areas, or was it merely using them as a bargaining chip? Did leaders offer to relinquish control over resource-endowed territories if their other demands were met? If aggressors demonstrated little interest in grabbing or holding oil or natural gas reservoirs, MIDs were not classic oil wars, even if they occurred in hydrocarbon-endowed territories. Collectively, these strategies enabled me to classify MIDs with confidence, particularly those that resulted in more than twenty-five battle deaths, which generated more extensive evidentiary trails.[17]

A New Typology

My investigation revealed no conflicts that cleanly qualify as classic oil wars. Instead, the 180 contests involving hydrocarbon-endowed territories either were fought for oil but resulted in fewer than twenty-five fatalities or were motivated predominantly by other issues. I refer to the former conflicts as oil spats. The latter group includes red herrings, oil campaigns, and oil gambits. Collectively, these four categories account for all 180 MIDs involving territories with known or prospective oil or natural gas resources, from 1912 to 2010.[18] The rest of the chapter elaborates on the four categories and their constituent MIDs, showing how each group supports the book's central argument: that states have little interest in fighting for oil resources.

Oil Spats

In October 2014, the Italian oil company ENI began surveying for gas resources off the island of Cyprus. Turkey, which has a long-standing claim to the island, responded by deploying a frigate to monitor ENI's activities. It also dispatched its own seismic survey ship, the *Barbaros*, to Cyprus's exclusive economic zone (EEZ). The Greek Cypriot government protested Turkey's actions and suspended talks on the disputed island's political reunification. Politicians and news outlets on both sides of the contest denounced their opponents; Turks and Turkish Cypriots lambasted Cyprus for its unilateral resource exploration in contested waters, while Greek Cypriots and their supporters in Athens retorted that Turkey had no right to interfere with activities in Cyprus's EEZ.[19] Yet, in spite of these rhetorical ripostes, the incident failed to escalate. When the *Barbaros* withdrew in April 2015, tensions declined, enabling reunification talks to resume the next month.

This episode was an oil spat: a minor militarized confrontation inspired largely by states' petroleum ambitions. My analysis identified nineteen of these incidents, listed in table 4.2. As the table indicates, oil spats have occurred in many geographic regions and have been prosecuted by a wide variety of states. The table also reveals that most of the oil spats occurred after the first energy crisis (1973–1974), when oil prices spiked.[20] This timing suggests that states' willingness to fight for oil is, to some degree, connected to the resource's economic value, as oil war believers have asserted. However, since these episodes were universally mild, this inclination evidently remains highly circumscribed, even when oil's value rises.

Like the 2014 incident between Cyprus and Turkey, most oil spats in the 1912–2010 time period revolved around hydrocarbon exploration and development. In these spats, one country attempted to conduct seismic surveys or drill for oil or natural gas resources and the other state tried to disrupt those efforts, through threats or minor shows of force. In the mid-1970s, Libya and Tunisia sparred over drilling platforms in the contested waters of the Gulf of Gabès.[21] Greece and Turkey have engaged in four oil spats, precipitated by their efforts to develop

TABLE 4.2 Oil spats

PARTICIPANTS	CONTESTED HYDROCARBON-ENDOWED TERRITORIES	MID DATES (ID NUMBER)
Argentina–Chile	Strait of Magellan	1980–81 (2083)
Argentina–United Kingdom	Falkland/Malvinas Islands	1976 (363)
Argentina–Uruguay	Rio de la Plata estuary	1969 (1172)
Azerbaijan–Iran	Caspian Sea	2001 (4317)
China–Vietnam	Gulf of Tonkin; Spratly Islands; Vanguard Bank	1993 (4029); 1994 (4030); 1998 (4328)
Eritrea–Yemen	Hanish Islands	1995–96 (4121)
Ethiopia–Somalia	Ogaden region	1974 (1427)
Germany–Romania	Romania	1940–41 (3825)
Greece–Turkey	Aegean Sea	1974 (1292); 1976 (1289); 1981–82 (2175); 1986–87 (2179)
Guyana–Venezuela	Essequibo Province	1999 (4260)
Indonesia–Vietnam	Natuna Islands	1980 (3610)
Iran–Iraq	Border zone	2009–10 (4546)
Libya–Tunisia	Gulf of Gabès	1977 (3014)
Turkey–United Kingdom	Mosul Province	1925–26 (3185)

contested petroleum resources in the northern Aegean Sea.[22] In the 1990s, China and Vietnam sparred over oil exploitation in the Gulf of Tonkin and near Vanguard Reef, in the Wan-an Bei 21/Tu Chinh exploration block.[23] In 1999, Venezuela's president, Hugo Chavez, threatened Guyana after the state began issuing oil exploration licenses for the continental shelf off the states' contested Essequibo Province. A few days later, Guyana reported troop movements and airspace violations, as well as a Venezuelan garrison firing shots along the disputed border.[24]

Oil spats often precipitate hyperbolic rhetoric while they are under way. In both participant states, the local press whips up nationalist sentiment. Foreign journalists predict that the incident will escalate into a larger conflict. Domestic populations launch demonstrations, denouncing their adversary and encouraging a robust state response. Government officials lodge protests with international organizations, such as the UN. Opposition politicians seize the opportunity to excoriate state leaders for their insufficient bellicosity. Collectively, these responses make oil spats seem significant and threatening, since it appears highly likely that they will escalate, potentially spiraling into much larger conflicts.

Yet the militarized activities that occur during oil spats are consistently minor. Out of the nineteen historical spats, only two were fatal. In one of the exceptions, a Greco–Turkish MID (1986–1987), fatalities were restricted to the mainland, far from hydrocarbon resources, and occurred over a month before the Aegean component of the contest intensified.[25] Hence, the only truly petroleum-driven fatalities in all of the oil spats took place in Eritrea and Yemen's 1995–1996 confrontation over the Hanish Islands in the Red Sea, which killed approximately a dozen soldiers.[26]

In addition to producing few fatalities, oil spats fail to escalate. None of the nineteen incidents spiraled into a larger conflict.[27] Instead, leaders were quick to contain petroleum sparring, reining in the actors that initiated the confrontations and refraining from further military operations. Often, in the wake of oil spats, belligerents pledged to cooperatively manage contested oil and natural gas reservoirs or to resolve their long-standing territorial disagreements. On occasion, they responded to these incidents by referring their territorial disputes to the International Court of Justice or other international institutions for mediation. In addition, oil spats have little ability to inspire aggression indirectly by heightening bilateral acrimony and increasing states' willingness to fight later, over other issues. Only two of the historical oil spats, between Argentina and the United Kingdom (1976) and Ethiopia and Somalia (1974), were eventually followed by a major interstate conflict.[28] Moreover, as chapter 6 will show, connections between the Anglo–Argentine oil spat and the countries' subsequent Falklands War were minimal.

The rarity of oil spats and their relatively benign character offer strong evidence of states' reluctance to fight for petroleum resources. This finding is further supported by oil spats' common characteristics, which indicate that countries only engage in these confrontations under highly favorable circumstances: when costs are low or they anticipate substantial nonoil benefits from international aggression, in addition to petroleum payoffs. The first of these circumstances is the dispute scenario discussed in chapter 3. All nineteen historical oil spats occurred in dispute scenarios, rather than conquest scenarios. Sovereignty in each contested area was uncertain, so all of the spat's participants could legitimately claim the hydrocarbon-endowed territory. Thus, rather than "stealing" foreign oil, each aggressor was merely advancing or defending a potentially legal resource claim. Consequently, the aggressor could anticipate relatively muted local and international resistance, especially if it refrained from major military operations.

Second, most of the oil spats involved offshore hydrocarbon resources.[29] Conflicts over offshore reservoirs entail fewer obstacles than conflicts over onshore deposits, partly because of the ambiguity of maritime boundaries. Whereas most of the world's land borders have been settled, fewer than half of international maritime boundaries have been fully delimited. Thus, a large percentage of offshore hydrocarbon resources remain in disputed territories, reducing the intensity of local and international opposition to aggression. Occupation obstacles are also low in contests over offshore hydrocarbons, since the local population consists of island residents, at most, and is nonexistent in purely maritime conflicts. In addition, invasion obstacles are relatively limited, as offshore reservoirs tend to be less developed than onshore reservoirs, particularly in disputed territories. Accordingly, there is little petroleum infrastructure for international aggression to damage.

Third, three-quarters of the oil spats occurred between states that shared a history of militarized hostility that predated oil expectations or discoveries; they fought over other issues before they fought for control over oil.[30] A contentious bilateral history increases the benefits of petroleum sparring, as participants can advance other interests, including buttressing national pride, undermining a historical rival, and securing contested territories' other economic, symbolic, and strategic assets, in addition to pursuing their oil ambitions. Oil spats may also improve an aggressor's bargaining position in ongoing diplomatic negotiations and agreements on policy issues, as well as enhancing a leader's domestic popularity if her population is emotionally or materially invested in the dispute.[31]

Fourth, all of the oil spats were prosecuted by neighbors; states do not engage in long-distance petroleum sparring far from their home territories.[32] Geographic proximity increases oil spats' appeal by reducing the costs of contention; it is far easier to project power against a neighboring state than a distant adversary. Neighboring states are also more likely to be engaged in ongoing territorial disputes.[33]

Altogether, oil spats' common characteristics indicate that states are highly selective, even when it comes to minor acts of petroleum-oriented aggression. They pick the most opportune moments to launch their oil spats: when territories are contested, resources are offshore, and their adversary is a state with which they share a history of hostility. Moreover, most of the time, countries still refrain from sparring for petroleum. Even oil spats are apparently rarely worth the effort.

Red Herrings

Most of the conflicts involving hydrocarbon-endowed territories were not fought for oil resources. Over 85 percent of the 180 MIDs that I examined closely were red herrings. In these conflicts, aggressors were fighting over territories that contained known or prospective oil or natural gas resources. However, hydrocarbon ambitions were not a significant motive for their military actions. In some of the red herrings, aggressors had no petroleum aspirations. In others, oil or gas ambitions were a minor, additional motive for their international attacks.[34] In addition, most red herrings resulted in fewer than twenty-five battle deaths, making them doubly divorced from classic oil wars; they were mild episodes fought for other reasons.

To discuss the red herrings, I focus on the conflicts that resulted in twenty-five or more battle deaths. These thirty-nine cases, listed in table 4.3, merit additional attention, since they are the potential classic oil wars: severe militarized interstate conflicts aimed at acquiring petroleum resources.[35] In addition, previous scholars have attached the classic oil war label to many of these cases, including the Chaco War, the Iran–Iraq War, the Falklands War, Egypt and Israel's conflicts over the Sinai Peninsula, Nigeria and Cameroon's Bakassi Peninsula confrontations, and China and Vietnam's 1974 clash over the Paracel Islands.[36] My analysis challenges those interpretations by finding that petroleum aspirations

TABLE 4.3 Severe red herrings

PARTICIPANTS	CONTESTED HYDROCARBON-ENDOWED TERRITORIES	MID DATES (ID NUMBER)
Albania–Yugoslavia	Albania	1949–51 (2328)
Argentina–United Kingdom	Falkland/Malvinas Islands	1982 (3630)
Bolivia–Paraguay	Chaco Boreal	1931–35 (1027)
Bosnia and Herzegovina; Croatia; Yugoslavia	Bosnia and Herzegovina; Croatia	1992 (3555); 1992–93 (3557); 1992 (3556); 1993–94 (4341)
Burkina Faso–Mali	Agacher Strip	1985–86 (2583)
Cameroon–Nigeria	Bakassi Peninsula	1995 (4165); 1996–97 (4166)
Chad–Libya	Aouzou Strip; Chad	1977–80 (3631)
China–Japan	Manchuria	1931–33 (129); 1937–41 (157)
China–Taiwan	China; Taiwan	1949–50 (633); 1953–56 (50); 1958 (173)
China–Vietnam	Paracel Islands; Spratly Islands	1974 (355); 1988 (2749)
Egypt–Israel	Sinai Peninsula	1966–67 (1035); 1967–70 (1480); 1971–73 (1046)
Ethiopia–Somalia	Ogaden region	1934–36 (111); 1963–64 (1425); 1977–78 (2069); 1978–79 (2070); 1980–81 (2071); 1982–83 (2072); 1984–85 (2074); 1985 (2075)
Indonesia–Malaysia	Brunei; Sarawak	1963–65 (1070)
Indonesia–Netherlands	Western New Guinea	1953 (2000); 1960–62 (1021)
Iran–Iraq	Border zones; Khuzestan Province	1979–80 (2114); 1980–88 (2115)
World War II	Albania; Austria; Czechoslovakia; Denmark; France; Germany; Hungary; Italy; Manchuria; Netherlands; Poland; Yugoslavia	1939–45 (258); 1939 (108); 1939 (169); 1940 (3706); 1940–41 (3813)

were, at most, a marginal incentive for aggression in each of the thirty-nine cases. Rather than fighting for territories' hydrocarbon resources, aggressors were predominantly motivated by other interests: aspirations to regional hegemony or political independence, national security concerns, domestic politics, national pride, windows of opportunity generated by changing regional power balances, and disputed territories' other economic, strategic, and symbolic assets. To elaborate on these findings, I briefly discuss each of the alternative motives, showing how they inspired aggression in the severe red herrings.[37]

Domestic politics drove Cameroon and Nigeria's severe confrontations over the Bakassi Peninsula, as well as Burkina Faso and Mali's over the Agacher Strip. In the former case, Cameroon and Nigeria had contested control over the purportedly oil-rich peninsula since the 1970s. However, the dispute escalated in the 1990s because Nigeria's president, Sani Abacha, faced intense domestic opposition and wanted to shift attention away from his regime's political and economic failings. He invaded the Bakassi Peninsula to generate a rally-'round-the-flag effect.[38] Similarly, the "Christmas War" between Burkina Faso and Mali (1985–1986) occurred in the long-contested Agacher Strip. Both states believed that the territory contained rich mineral endowments. However, the conflict broke out when the increasingly unpopular Malian president, Moussa Traoré, seized on a controversial Burkinabe census in the disputed region as an excuse to launch an international attack, which aimed to divert popular attention from internal unrest.[39] As chapters 5 and 6 will show, domestic politics also contributed significantly to the initiation of the Chaco and Falklands Wars. In each of these contests, aggressors had little to no interest in contested territories' petroleum resources.

In other severe red herrings, national security concerns dwarfed petroleum ambitions. When Israel seized the Sinai Peninsula during the Six-Day War (1967), it gained control over Egyptian oil fields near Belayim and Abu Rudeis. These resources increased Israel's reluctance to withdraw from the peninsula after occupying it; the state's eventual departure (1975–1979) was conditioned on the United States guaranteeing its access to oil resources.[40] However, the invasion itself was motivated by national security concerns. Israelis wanted to preempt an Arab attack and create a buffer between themselves and Egypt. Likewise, national security interests dominated Egypt's efforts to retake Sinai; the Egyptians were more interested in repelling Israeli forces than in reclaiming oil fields.[41] As chapters 5 and 8 will show, the Iran–Iraq War and Iraq's invasion of Kuwait were also motivated predominantly by national security concerns; Saddam Hussein believed that Iran's revolutionary Islamist regime and, later, the United States threatened his government's survival.

In a number of the severe red herrings, aggressors were primarily bidding for regional hegemony. In the late 1970s, Libyan leader Muammar Qaddafi aimed to displace Egyptian president Anwar Sadat as North Africa's political leader and pursue his vision of pan-Arab and pan-Islamic unity. To advance these goals, he intervened in Chad's civil war, sending troops as far south as N'djamena and Lake Chad, in order to strengthen his influence in the neighboring country. Small amounts of oil were being produced near the lake at the time of Qaddafi's invasion. However, neither those nor the contested Aouzou Strip's purported petroleum endowments were a significant Libyan target.[42] In the Horn of Africa, clashes over the Ogaden, the large triangular region of Ethiopia adjacent to present-day Somalia, were also motivated by hegemonic ambitions. In the 1930s, Benito Mussolini aimed to expand Italy's colonial empire by seizing Abyssinia (present-day Ethiopia). Since the 1960s, a number of Somali leaders have aspired to establish "Greater Somalia," encompassing all territories inhabited by ethnic Somalis, including the Ogaden.[43] All of these aggressors were aware of the area's prospective petroleum endowments. However, oil ambitions were not a central motive for their attacks.

Lastly, the two world wars were provoked by Germany's and Japan's hegemonic aspirations in Europe and East Asia. As the next section and chapter 7 will discuss, these conflicts included campaigns that targeted petroleum resources. However, Japan and Germany did not initiate the world wars to grab foreign oil. In addition, the other severe conflicts that occurred in conjunction with the European War were not fought for oil resources, although some of them occurred in hydrocarbon-endowed territories, including Albania, Poland, and Yugoslavia. Instead, in those severe red herrings, aggressors took advantage of the permissive conditions created by Germany's expansionism to advance their own longstanding territorial claims, often against historical rivals.

Both of China and Vietnam's severe clashes in the South China Sea were triggered by shifting regional power balances linked to bids for regional hegemony. China has aspired to increase its influence in the South China Sea since the late 1960s. However, the state's initially limited naval capabilities forced it to wait for windows of opportunity to advance this goal. One window opened in late 1973 around the Paracel Islands, which China contested with Vietnam. The South Vietnamese government, preoccupied with its ongoing war with North Vietnam, had reduced its presence in the disputed archipelago. The United States had also withdrawn most of its forces from South Vietnam and had little interest in reengaging in the region.[44] Meanwhile, Beijing worried that Soviet activity in the area would increase following North Vietnam's anticipated victory over South Vietnam. Before this could happen, China seized the opportunity to capture full control over the islands in January 1974.[45] At the time, it was commonly believed that the area around the Paracels contained petroleum resources.[46] However, China had little interest in exploiting them, as it still produced abundant oil from its mainland territories and most of the islands' continental shelf was technologically inaccessible because of extreme water depths.[47]

China's second window of opportunity in the South China Sea opened in the late 1980s, as Sino–Soviet relations improved and the Soviet Union reduced its presence in the region. As a result, Beijing could finally occupy some of the reefs and cays it had claimed in the Spratly archipelago.[48] In March 1988, China's island initiatives triggered a clash with the Vietnamese navy near Johnson Reef. The People's Liberation Army Navy sank three Vietnamese ships, killing seventy-five sailors.[49] This deadly incident occurred in a region that, like the Paracels, was believed to contain valuable oil and gas deposits. However, the Sino–Vietnamese clash occurred in the midst of a global oil glut while China was still a net petroleum exporter, casting doubt on hydrocarbon explanations for the confrontation.

In many of the severe red herrings, aggressors were more interested in contested territories' other economic, strategic, or symbolic assets than their oil and gas endowments. Libya intervened in Chad's civil war partly to strengthen its control over the Aouzou Strip's uranium deposits, which could fuel Qaddafi's nuclear ambitions.[50] Burkina Faso and Mali were more attracted by the Agacher Strip's valuable manganese deposits than its rumored petroleum endowments.[51] China and Vietnam were both interested in the South China Sea's rich fisheries, and Beijing was particularly attracted by the area's strategic significance; the Paracel and Spratly Islands border critical sea lanes of communication, and Japan employed the latter as a launching pad for its aggression in the South Pacific during World War II.[52]

National pride contributed to many of the red herrings, including the Falklands War, the Chaco War, the Iran–Iraq War, Egypt and Israel's confrontations over the Sinai Peninsula, the Sino–Vietnamese clashes in the South China Sea, Ethiopia and Somalia's conflicts over the Ogaden, and World War II. National pride also underpinned the diversionary conflicts launched by Nigeria and Mali; without it, there would have been nothing for domestic populations to rally around.[53] Pride figured heavily in these conflicts because most of the belligerents shared histories of territorial competition and militarized hostility, which predated oil expectations or discoveries. These states had plenty to fight over, with or without oil.

Lastly, a substantial number of severe red herrings were attempts to obtain—or sustain—political independence. In 1918, Estonia prosecuted a successful war of independence against the Soviet Union. From 1949 to 1951, Albania attempted to maintain its independence from Yugoslavia. Following the Chinese Civil War, the Communists and Nationalists fought repeatedly for control over mainland China and Taiwan. From the 1950s to the 1960s, Indonesia challenged the Dutch in western New Guinea because the Netherlands had maintained its hold on the territory, which Jakarta believed it should control, after granting Indonesia independence in 1949. The Sukarno government also attempted to block the formation of the Federation of Malaysia and, after the state was successfully created from the territories of Malaya, North Borneo, and Sarawak in 1963, Jakarta attempted to undermine it through covert interventions and direct attacks. In the 1990s, Bosnia and Herzegovina and Croatia fought for independence from Yugoslavia. There were known or prospective oil resources at stake in each of these conflicts: in Albania, Estonia, mainland China, Taiwan, western New Guinea, Sarawak, Croatia, and Bosnia and Herzegovina.[54] However, the resources were not significant casus belli for any of the belligerents.

All in all, the red herrings offer little support for classic oil war claims. At most, oil and natural gas resources were a marginal added incentive for aggression in conflicts launched predominantly for other reasons. In many of the cases, petroleum ambitions played no causal role. The participants in red herrings evidently did not believe that oil was worth fighting for.

Oil Campaigns

On rare occasions, countries have launched major international attacks targeting oil resources. All but one of these conflicts, however, were oil campaigns. In oil campaigns, an aggressor's primary goal is to seize foreign petroleum resources.[55] Oil campaigns also kill hundreds, if not thousands, of combatants. It is therefore understandable that previous researchers have labeled most historical oil campaigns classic oil wars.[56] However, in doing so, they conflate oil's ability to inspire international conflict with its capacity to influence the trajectories of wars that are already under way. All of the historical oil campaigns occurred in the midst of major, ongoing conflicts that were not themselves caused by oil ambitions. Absent these existing wars, oil campaigners would have refrained from fighting for petroleum resources.

My study identified five historical oil campaigns, all of which occurred during World War I, the Second Sino–Japanese War (1937–1945), or World War II. These campaigns are Japan's invasion of the Dutch East Indies and British Borneo (1941–1942); Germany's attacks against the Soviet Union in World War II (1941–1942); Turkey and Germany's race to seize Baku, in the Soviet Caucasus, in the final year of World War I (1918); Germany's invasion of Romania in the same conflict (1916); and the United Kingdom's invasion of Mosul Province, in present-day Iraq, during the war's closing days (1918).[57] Each of these campaigns was largely driven by oil ambitions. The Japanese sought to obtain Southeast Asian petroleum resources. Germany aspired to grab Soviet oil fields during World War II, and in 1918, it competed with Turkey for this prize. In September 1916, after Romania declared war on Austria–Hungary, Germany launched a counterattack that gave it control over the state's rich oil fields at Ploieşti.[58] At the end of World War I, British troops seized prospective petroleum resources in northern Mesopotamia.[59]

Yet while each of these campaigns targeted oil, none of them provide compelling evidence that countries' desire to obtain petroleum resources is a significant cause of international conflict. Instead, each of these assaults occurred in the midst of an ongoing war. Japan attacked Southeast Asia during the Second Sino–Japanese War, also known as the China Incident. Germany initiated its Soviet campaigns well after each world war was under way. Turkey and the United Kingdom did not launch their attacks on Baku and Mosul until the waning months of World War I. The broader wars in which aggressors prosecuted their oil campaigns were not driven by petroleum ambitions. Instead, they arose from Germany's and Japan's hegemonic aspirations in Europe and East Asia.

The ongoing conflicts shifted participants' strategic calculations. They heightened national petroleum consumption, as belligerents required enormous amounts of oil-based fuels to power their war machines. They also constrained future oil campaigners' ability to purchase crude oil and petroleum products, as all of them were subjected to blockades or trade embargos. The United States began restricting Japan's oil imports in the late 1930s in response to the state's aggression in China and Southeast Asia. Germany was blockaded in both world wars. During the first world war, Germany's U-boat campaign disrupted British oil supplies, and the Entente's blockade of the Central Powers caused Turkish petroleum shortages. Eventually, each oil campaigner concluded that seizing foreign oil resources offered the only remaining, viable means of satisfying its wartime petroleum needs.

The ongoing wars therefore generated strong incentives for new acts of international aggression, targeting oil resources. They also created exceptional opportunities for these attacks. Since belligerents were already at war, they could expand the scope of contention more freely, particularly against targets that were already weakened by the ongoing conflict. Oil campaigners exploited these permissive conditions to obtain foreign resources. However, the same petroleum deposits failed to inspire oil grabs before these larger conflicts began. Moreover, as chapter 7 will show, even in wartime, aggressors refrained from launching oil campaigns until they had exhausted all other means of satisfying national petroleum needs, including developing domestic resource endowments, producing synthetic fuels, acquiring overseas concessions, increasing foreign trade, and intensifying international diplomacy.

The five historical oil campaigns therefore underline petroleum's exceptional importance for modern warfare. They also demonstrate that oil needs shape belligerents' wartime behaviors. However, they do not indicate that oil is a significant cause of international conflict. Instead, oil campaigns occurred in the midst of ongoing international conflicts that had already started for other reasons. These campaigns were also initiated as a last resort. States had little interest in fighting for petroleum resources.

Oil Gambit

Of the 180 MIDs involving hydrocarbon-endowed areas, only one was an oil gambit: Iraq's invasion of Kuwait. Saddam Hussein launched this severe international attack, targeting oil, in the absence of an ongoing war. Unsurprisingly, the Iraqi case tops most lists of classic oil wars; Jan Selby calls it "an oil war, if ever there was one."[60] Yet labeling the conflict a classic oil war is an oversimplification. Iraq was attempting to seize Kuwait's oil fields. However, as chapter 8 will demonstrate, depictions of this conflict as a straightforward, greedy oil grab are mistaken. More nuanced interpretations of the invasion that emphasize Iraq's oil needs are also incomplete. Instead, the fundamental motive for Saddam's aggression was his belief that the United States was determined to overthrow his regime, coupled with his eventual conclusion that conquering Kuwait offered the only viable means of escaping the American threat. Absent this broader security concern, Saddam would not have attacked Kuwait. I label this type of conflict an oil gambit to signify that, although the aggressor targeted petroleum, it was less interested in grabbing the resource than in achieving a larger, political goal.[61] Oil gambits are instrumental.

They are also desperate. Saddam, like the oil campaigners, launched his oil gambit as a strategy of last resort after exhausting all other means of sustaining his regime's survival in the face of intensifying internal and international threats. This finding suggests that, if we decide to label Iraq's invasion a classic oil war—a plausible choice, based on its severity and target—we also need to recognize that these conflicts look quite different from what most of us have imagined. Saddam did not greedily grab his neighbor's oil. Nor was he acting solely to satisfy his state's petroleum needs. Instead, in this unique historical episode of severe, oil-driven international aggression, the Iraqi leader believed that he was confronting a broader existential threat.

Deepening the Analysis

Having outlined the core characteristics of each of the four types of contention that actually occur in hydrocarbon-endowed territories, the book's remaining chapters examine cases from each of the categories to elucidate how oil ambitions do—or do not—contribute to interstate conflict. My case selection was based on the criteria of representativeness and importance. Conflicts had to accurately illustrate their particular analytic category. However, I also prioritized cases that are commonly identified as classic oil wars, in order to reevaluate petroleum's contribution to each conflict. Chapter 5 therefore presents two red herrings that are regularly described as oil-driven: the Chaco War between Bolivia and Paraguay and the Iran–Iraq War. Chapter 6 examines an oil spat: Argentina and the United Kingdom's 1976 confrontation near the Falkland/Malvinas Islands. It also discusses the Falklands War, demonstrating that, while some previous authors have labeled it a classic oil war, it was actually another red herring. Chapter 7 explores Japan's and Germany's oil campaigns during World War II, and chapter 8 assesses the sole oil gambit, Iraq's invasion of Kuwait. Chapter 8 also includes a postscript establishing that the United States' invasion of Iraq was not a classic oil war. Collectively, these case studies reinforce this chapter's central finding: that states avoid fighting for oil resources.

5

RED HERRINGS

The Chaco and Iran–Iraq Wars

The Chaco War (1932–1935) and Iran–Iraq War (1980–1988) appear on many lists of classic oil wars. In the former conflict, Bolivia and Paraguay were supposedly fighting for control over the Chaco Boreal's prospective petroleum resources, possibly at the behest of two multinational oil companies: Standard Oil and Royal Dutch Shell. In the latter, Saddam Hussein purportedly attacked his neighbor partly to annex the oil-rich province of Khuzestan, thereby expanding Iraq's petroleum resources and revenue. Both of these interpretations are widely accepted. Yet neither is correct. These conflicts were not classic oil wars, where petroleum ambitions are a leading incentive for international aggression. Instead, they were red herrings; the desire to grab more oil resources did not motivate any of the belligerents.

This chapter focuses on the Chaco and Iran–Iraq Wars because these are two of the deadliest historical red herrings, as well as the most broadly misinterpreted. The Chaco War caused over 90,000 fatalities out of the belligerents' combined populations of fewer than five million. The Iran–Iraq War resulted in over 350,000 battle deaths. If petroleum did not cause these severe interstate conflicts, some of the oil wars myth's most striking supporting evidence vanishes.

Since these red herrings were not caused by oil ambitions, the four sets of obstacles described in chapter 3 have limited bearing on them. The Bolivian and Paraguayan governments were not contemplating invasion, occupation, international, and investment impediments before the Chaco War because they did not believe that the contested region contained petroleum resources. Oil-related obstacles to aggression were therefore irrelevant to their decision making. The Iraqi government, in contrast, was attacking a major oil-producing province. Consequently, authorities were more attentive to these impediments and concern about occupation obstacles, in particular, shaped their aggression in Khuzestan. However, since the Iraqis were not fighting to grab Iranian oil resources, petroleum-related impediments did not deter their invasion.

Rather than being driven by oil ambitions, the Chaco and Iran–Iraq Wars, like many other red herrings, were motivated by a combination of national security concerns, domestic politics, and national pride: issues the governments believed were worth fighting for. In addition, belligerents in the former contest were concerned with petroleum transportation; Bolivians wanted to build an oil pipeline across the Chaco Boreal to the Río Paraguay. However, as previous oil war scholars have noted, efforts to secure petroleum transit routes are not classic oil wars. More importantly, in both of these conflicts, transportation concerns were secondary to states' other motives for international aggression.

As the following case studies demonstrate, the oil war label was attached to the Chaco and Iran–Iraq conflicts not because it accurately reflected participants' conflict goals but by mistake or for strategic reasons. Belligerents were not prosecuting classic oil wars. However, many people found it convenient to accuse them of it. The case studies present the classic oil war interpretations of each red herring, explain why the explanations emerged, show that each is inaccurate, and identify the other issues that states were fighting for instead of oil.

The Chaco War (1932–1935)

The Chaco Boreal is a large, roughly circular territory located in the center of South America, where Bolivia's southeast meets Paraguay's northwest. Approximately one hundred thousand square miles in size, the Chaco is bounded to the east by the Río Paraguay, to the south by the Río Pilcomayo, and to the west by the Andean foothills. Described as a "green hell" by nineteenth-century travelers, it is an inhospitable area, consisting predominantly of dry scrublands. The territory contains few perennial surface water sources so, during the dry season, from May to October, the Chaco is parched. During the wet season, from November to April, much of the region floods, creating muddy swamps that impede transportation and breed insect-borne diseases.[1]

In the early twentieth century, sovereignty over the Chaco Boreal was ambiguous. Spanish colonial authorities failed to delimit a border between Bolivia and Paraguay before the states' independence, and the records and maps officials left behind were incomplete and inconsistent. Both states therefore possessed legitimate claims to the region. Their dispute emerged in 1852, when Argentina and

MAP 5.1. The Chaco Boreal (1932)

Paraguay signed an agreement stating that the Río Paraguay belonged entirely to the latter. Bolivia protested, asserting its own rights to the waterway.[2] A quarter century later, Bolivia and Paraguay initiated formal negotiations over the Chaco Boreal. Over the next four decades, they signed numerous treaties concerning the region. However, none of these agreements were ratified by both states' legislatures, so the dispute remained unresolved.[3]

There was limited economic rationale for the states' attachments to the Chaco. Bolivia made few efforts to develop the territory before the 1932 war. The country's economic and political hubs were located in the Andean highlands, in the cities of La Paz and Cochabamba and mining centers like Potosí. These areas, in the Altiplano, were poorly connected to the eastern provinces that bordered the Chaco. Consequently, there was little incentive for Bolivia to develop the contested territory, as transporting supplies to the area would be costly and there were no outlets for local production. By 1920, the only Bolivian settlements in the region were rudimentary military *fortínes* (forts), housing small numbers of soldiers, along the Río Pilcomayo.[4]

Paraguay, in contrast, had stronger geographic connections to the Chaco and made a greater effort to develop it. In the late nineteenth century, the government sold off large tracts of Chaco public lands to bolster state revenue. Argentinean investors, who were the primary beneficiaries of the sales, used their newly acquired territories for cattle ranching. They also exploited the Chaco's indigenous *quebracho* trees, which were a natural source of tannins and lumber. By 1932, the Paraguayan government claimed, the Chaco was supplying one-third of the state's income.[5] However, these economic activities were still concentrated near the western bank of the Río Paraguay. The only nonindigenous people to penetrate the Chaco's interior were Mennonite settlers who began arriving in the territory in the 1920s, with the Paraguayan government's encouragement.[6] Thus, when the war began, many parts of the Chaco were not under the effective authority of either claimant state.

No oil development occurred in the Chaco before the war. However, Bolivia possessed an active petroleum industry in the Andean foothills along the Chaco's western edge, between the towns of Santa Cruz and Tarija. La Paz issued concessions for the area around the turn of the century and prospectors drilled the region's first oil well in 1911. However, the industry's early development was slowed by a lack of indigenous capital, the area's geographic isolation, and limited local petroleum demand. It was only in the 1920s, when the Standard Oil Company of New Jersey acquired concessions in the region, that development accelerated. Standard Oil of Bolivia, the company's local subsidiary, drilled approximately thirty wells before the war, striking oil near Bermejo in 1922. Standard also built two small refineries, at Camiri and Sanandita, to produce petroleum products.[7]

Yet Standard's Bolivian output was low. Before the war, the company extracted fewer than 150 barrels of oil per day: only enough to supply its local operations. In the early 1920s, Standard had little incentive to increase its output, because a global oil glut had reduced petroleum prices. These failed to recover in the mid-1920s and fell further during the Great Depression.[8] Standard was also deterred from increasing production by a lack of outlets for its oil. At the time, it was impossible to move crude oil or petroleum products from the Andean foothills to the Altiplano. There were no rail links between Santa Cruz and Cochabamba and transporting the resources via pipeline was physically impractical because of a seven-thousand-foot altitude gain.[9] Rail linkages to Argentina were more promising, especially after 1922, when Bolivia and Argentina agreed to extend the Formosa–Embarcación line to Yacuiba and Santa Cruz. However, construction proceeded slowly, as a result of misgivings in both states.[10] In addition, in 1927,

Argentina increased tariffs on rail transportation of Bolivian oil. Buenos Aires also repeatedly refused La Paz's requests to construct a pipeline across its northern territory, which would allow Bolivia to move petroleum quickly and cheaply from its oil fields to a port on the Río Paraná.[11] For the moment, Bolivia's oil was trapped.

Meanwhile, competition over the Chaco was mounting. During Paraguay's civil war (1922–1923), Bolivia had seized the opportunity to intensify its construction of fortínes, extending its Chaco outposts progressively farther down the Río Pilcomayo. By 1927, the state had also begun to spread its installations northward, toward Santa Cruz. Paraguay responded by expanding its own fortínes network.[12] The increased concentration of military forces, combined with rising nationalist sentiment and diminishing prospects of a negotiated dispute settlement, heightened the potential for an international clash.

The states' first fatal Chaco confrontation occurred in February 1927. Bolivian troops, stationed near Fortín Sorpresa, captured and killed a Paraguayan soldier, Adolfo Rojas Silva, son of the former Paraguayan president Liberato Marcial Rojas. Rojas Silva's death sparked state-sanctioned protests throughout Paraguay; fifty thousand people demonstrated in Asunción alone.[13] Tensions escalated further in early December 1928, when Paraguayan troops occupied Bolivia's Fortín Vanguardia. Bolivian soldiers retook the fort a few days later and subsequently attacked two Paraguayan fortínes, Boquerón and Mariscal López. These incidents "inflamed public passion to a dangerous degree" in both states; only foreign diplomatic intervention prevented the confrontation from escalating.[14] In September 1931, Bolivian forces raided Paraguay's Fortín Samaklay. The Paraguayan government attempted to cover up the loss, provoking riots in Asunción when the deceit was discovered. By the end of the year, "rumors of impending war were rampant."[15]

The incident that finally tipped the states into war occurred the next year. On June 14, 1932, a Bolivian expeditionary force led by Major Oscar Moscoso Gutiérrez arrived at Laguna Chuquisaca/Pitiantuta.[16] The Bolivians chased off a contingent of Paraguayan soldiers at Fortín Carlos Antonio López and established their own post on the lake. Paraguayan forces retaliated by seizing the Bolivian fort on July 16. Third parties again attempted to restrain the states in order to prevent the confrontations from escalating into outright war. However, Bolivia's president, Daniel Salamanca, refused to back down. Labeling the Paraguayan counterattack "a new aggression against the dignity of Bolivia," he directed his forces to seize the Paraguayan fortínes of Boquerón, Corrales, and Toledo.[17] With these new attacks, the Chaco War was under way. It would continue for three years, becoming Latin America's deadliest twentieth-century conflict.[18]

Oil war explanations for the Chaco conflict emerged soon after the war began. Initially, leftist intellectuals in Bolivia, Paraguay, and Argentina blamed the contest on Standard Oil, claiming that the company had driven the Bolivian government to war in order to seize the Chaco Boreal's prospective petroleum resources.[19] By the end of 1934, these accusations had broadened to indict another oil company; believers claimed that Royal Dutch Shell was supporting Paraguay, causing the belligerents to fight a proxy war on behalf of the two multinationals.[20] Both of these arguments are regularly repeated by contemporary classic oil war believers.[21] However, at the time, the initial version attracted more popular attention, largely because of its promotion by Louisiana senator Huey P. Long.

In multiple speeches on the floor of the US Senate from May 1934 to January 1935, Long blamed the Chaco War on Standard Oil. As he put it, "The Bolivian and Paraguayan Governments are now engaged in war as a result of the agitation for concessions granted by the Bolivian Government to the Standard Oil Co. of New Jersey." Long asserted that "this Standard Oil Co. is financing the Chaco war, hoping to get two million four hundred and some odd thousand acres of territory" that belonged to Paraguay. As Long pointed out, in 1878, US president Rutherford B. Hayes had awarded the area between the Río Pilcomayo and the eastern portion of the Río Verde to Asunción through international arbitration. According to Long, Bolivia was therefore trampling on Paraguayan sovereignty and the United States' political authority, at Standard Oil's behest.[22]

Long's speeches circulated widely, and upon reading them, the Paraguayan government also embraced the oil war explanation. The state's 1934 submission to the League of Nations, which was investigating the Chaco dispute, extensively quoted Long's allegations.[23] The Paraguayan press, which had been accusing Standard of instigating the war for over a year, also lionized the senator, describing him as hero fighting for "justice and truth." In August 1934, the Paraguayan military rechristened a captured Bolivian outpost Fortín Senator Long.[24] "The Kingfish" had found new acolytes who persistently reiterated his oil war argument during and after the conflict.

Yet there was no oil in the Chaco Boreal. Geologists from Standard, who had explored the Andean foothills and the Chaco plain, concluded that oil fields were limited to the area west of the sixty-third meridian, between Santa Cruz and Tarija. As one report observed, "It is very doubtful that oil exists in a geological structure such as that of the Chaco."[25] Gordon Ireland, a prominent scholar of Latin American territorial disputes, observed in 1938 that there were "oil and minerals in the higher western part" of the Chaco, but not in the contested plains. Ronald Stuart Kain, describing the disputed territory in 1935, asserted that "no petroleum or other minerals have been discovered in the Chaco Boreal, nor is it expected that any will be found, in view of geologists' reports."[26] Following Long's speeches in the US Senate, one of his opponents observed, "I think it is admitted by everyone that there is no oil to be found in the Chaco."[27] Even Long eventually conceded, acknowledging in January 1935 that he had conflated Bolivia's oil fields and the contested Chaco territory.[28]

Standard Oil's behavior during the conflict was also inconsistent with classic oil war arguments. Rather than supporting the Bolivian government during the conflict, as one would expect if it sought additional concessions, the company refused the state's demands to increase petroleum production, as it was contractually obligated to do in wartime. Standard also falsely claimed that its refineries were unable to produce aviation gasoline and shipped drilling equipment and other supplies out of the country. When the Bolivian military appropriated many of the company's trucks for use in the war, Standard sued. The Bolivian government was so antagonized by the company's lack of cooperation that it seized its refineries in 1933, in the midst of the war, and expropriated all of its assets in 1937.[29]

The Bolivian and Paraguayan governments were also aware of the Chaco's limited oil prospects. Bolivia had received Standard's pessimistic prewar assessments and widely circulated a report highlighting the area's slim petroleum potential.[30] A Paraguayan military publication issued during the war observed that Bolivia's oil was in the Andean foothills, "en la immediate proximidad del Chaco Paraguay," but not in the Chaco itself.[31] Analyses of the dispute published before the war's outbreak in 1932 failed to mention oil resources as a source of territorial competition.[32] In addition, the states' leaders did not identify potential oil endowments as a cause of conflict at the war's outset. President Salamanca emphasized the need for better oil transportation in his public speeches and in discussions with his general staff. However, he did not suggest that the Chaco Boreal contained oil. Paraguay's president, Eusebio Ayala, initially "scoffed" at the claim that Standard Oil was driving the conflict.[33]

Because neither of the belligerents believed that the Chaco Boreal contained petroleum resources, authorities were not contemplating oil-related obstacles to aggression when deciding whether to fight over the contested territory; the invasion, occupation, international, and investment impediments discussed in chapter 3 did not enter into their thinking. Yet, if petroleum ambitions—and the obstacles to realizing them—did not influence leaders' decision making, why are oil war interpretations of the Chaco conflict so pervasive? Why was Long's rhetoric compelling in the 1930s, and why has the classic oil war explanation persisted over time? Finally, if Bolivia and Paraguay were not fighting over petroleum resources, why did they prosecute such a devastating war?

The classic oil war interpretation gained and maintained traction for a number of reasons. The first was geographic confusion. Observers of the war, especially outside the belligerent states, had little knowledge of the Chaco's geography. Hence, many assumed that Bolivia's oil fields, located in the Andean foothills, extended into the plain.[34] Long contributed to this confusion when he repeatedly asserted that Standard was pursuing oil concessions in the territories that the Hayes award had granted to Paraguay.[35] However, as Long's critics pointedly observed, that area was three hundred miles from Bolivia's oil fields. Unsurprisingly, given people's enthusiasm for oil war arguments, this geographic correction attracted less attention than Long's initial accusation. Many observers continued to assume that the contested Chaco Boreal possessed abundant petroleum resources, even after Long acknowledged the misrepresentation.

The second reason for the classic oil war interpretation's widespread adoption was its instrumental utility for belligerents and conflict observers. Long seized on the Chaco War because it presented an opportunity to lambast his nemesis: Standard Oil.[36] In the late 1920s, when Long was governor of Louisiana, Standard had stridently resisted his efforts to increase taxes on oil production, going so far as to sponsor an impeachment campaign against him. Although the attempt failed, from that point on, Long seized—or manufactured—any opportunity to attack the company. In his speeches on the Chaco War, Long accused Standard of "criminality, rapacity, and murder" and encouraged the US Senate "to seize this criminal, this culprit, this murderer."[37] The company's avaricious reputation, domestically and internationally, also made it an easy target. People were ready to accept that Standard could and would incite a classic oil war.

The Paraguayan government also latched onto the classic oil war explanation for instrumental purposes. By reiterating Long's claims, suggesting that La Paz was cooperating with Standard to perpetrate an oil grab, officials could discredit Bolivia's motives for war and curry international favor for Paraguay's cause. The Bolivian government employed the same tactic, accusing Paraguay of launching the war at the behest of "foreign capitalists" in order to seize Bolivia's existing oil fields.[38] Meanwhile, members of Bolivia's political opposition used the classic oil war argument to attack President Salamanca. By suggesting that he was in league with Standard, they could intensify populist hostility toward the regime. Lastly, as the war dragged on and losses mounted, more Bolivians embraced the classic oil war interpretation, as it offered some explanation for the ruinous conflict. That explanation may have been false. But it provided some psychological comfort by suggesting that there had been an underlying rationale for a war that seemed increasingly pointless.[39]

A third reason for the classic oil war argument's broad acceptance is conflict geography. The final phases of the Chaco War were prosecuted alongside Bolivia's oil fields. By late 1934, despite having the smaller military, Paraguay had outmatched Bolivia's army and seized most of the Chaco Boreal. In early 1935, Paraguayan forces attempted to seize Bolivia's refinery at Camiri and captured the Camatindi oil camp. In March, they approached oil wells at Cerro Teiguate.

However, Bolivian regiments, including one that had been renamed Defenders of the Oil, successfully fought them off. Petroleum ambitions did not drive these confrontations; Paraguay was attempting to terminate the war, and in the states' 1938 peace settlement, it returned all of the oil-endowed territories to Bolivia. However, the clashes' location, essentially at the base of Bolivian oil derricks, reinforced popular perceptions that the Chaco conflict was a classic oil war.[40]

Lastly, the classic oil war explanation gained traction because of Bolivian officials' concern with petroleum transportation. In the late 1920s, the state was eager to increase national oil output. The country was suffering economically, because of a significant downturn in the price of tin, its primary export commodity.[41] To compensate for the drop in revenue, the Siles (1926–1930) and Salamanca (1931–1935) governments pressed Standard Oil to increase petroleum production, with the goal of transforming their state into a major oil exporter.[42] However, Bolivian officials recognized that any attempts to export oil would be hampered by transportation impediments. Confronted with a physically untenable transit route through the Andes, expensive rail linkage to Argentina, and Buenos Aires' refusal to permit a pipeline across its northern territories, Bolivian authorities concluded that their only viable export option was a pipeline across the Chaco to a port on the Río Paraguay, located between Fuerte Olimpo and Bahía Negra.[43]

Unlike petroleum resources themselves, Bolivia's desire for an oil outlet contributed to the Chaco War's onset.[44] In a speech to the Bolivian Congress on August 6, 1932, following the confrontation at Laguna Chuquisaca/Pitiantuta, President Salamanca asserted,

> On the eastern slope of its mountains, Bolivia possesses great oil wealth, including some drilled wells that could be exploited immediately. We need these resources, yet are forced to view them as sterile wealth. Bolivia cannot transport its oil to Argentina because that country, out of self-interest, has blocked the way with harsh protectionist duties. The natural and logical remedy would be to construct a pipeline to the Río Paraguay. But the Republic of Paraguay is there, occupying Bolivian territory and blocking the path. Bolivia cannot resign itself to live miserably, as an isolated country, and must seek out the necessary conditions for its full life.[45]

Proponents of the oil war argument, including Long, highlighted Salamanca's speech as evidence of Bolivia's perfidious, petroleum-oriented intentions.[46]

However, as previous authors have noted, conflicts over oil transportation are conceptually distinct from classic oil wars.[47] Moreover, Bolivia's pipeline pursuits were more of a pipe dream than a practical goal. Standard Oil's engineers had investigated the Chaco pipeline idea in the early 1920s but abandoned it by 1925.[48] They recognized that the region's extreme environmental conditions would impede pipeline and port construction and interrupt resource flows. In addition, port sites above Fuerte Olimpo were ill suited to oil transportation. Most were flooded for parts of the year, and in the dry season, water depths dropped to less than 7 feet. The river was therefore unable to handle "even the smallest ocean-going tanker" of the time.[49] In addition, even if Bolivia could load oil supplies along the Río Paraguay, these would still have to travel through Argentina, via the Río Paraná, and Buenos Aires could deny them the use of transit facilities.[50] Shipments could also be intercepted as they passed through Paraguay, especially if the countries had recently fought a major war. The only way to stifle this resistance would be for Bolivia to thoroughly vanquish its neighbor: an outcome the Bolivian general staff deemed impossible.[51]

Salamanca's attachment to an oil export route via the Río Paraguay was therefore impractical. However, the idea possessed enormous symbolic resonance because of its connection to a larger, long-standing objective shared by most Bolivians: regaining sea access.[52] Bolivia had become a landlocked state when it lost its Litoral province to Chile in the War of the Pacific (1879–1883). Although the country's goods could still move freely to the coast, successive Bolivian governments considered that insufficient.[53] Thus, over the next fifty years, they persistently sought an independently controlled route to the Pacific. These hopes finally evaporated in 1929, when Chile and Peru formally divided the contested coastal regions of Tacna and Arica between themselves, offering no territory to Bolivia.[54] Since La Paz's efforts to develop an export route to the Atlantic via the Amazon had already floundered, a port on the Río Paraguay offered the only remaining means of overcoming Bolivia's "geographical asphyxia."[55] Consequently, much of the population had embraced the idea, for largely psychological reasons, well before oil transit concerns entered the picture.[56]

In addition to Bolivians' fixation on an ocean outlet, three other factors, unconnected to oil resources or transportation, encouraged the Chaco War's outbreak. The first was both belligerents' fears of additional territorial dismemberment.[57] In addition to losing its Litoral province in the 1880s, Bolivia lost its rubber-rich Acre province to Brazil in the Treaty of Petropolis (1903). In the War of the Triple Alliance (1864–1870), Paraguay ceded territories north of the Río Apa to Brazil and the Chaco Central, a region south of the Río Pilcomayo and north of the Río Bermejo, to Argentina. Both states were therefore loath to relinquish more land. Paraguay, in particular, would lose half of its territory, as well as a third of state revenue, if Bolivia seized the entire Chaco Boreal. Asunción was therefore in a "fight for survival."[58] Bolivia, in contrast, did not face an existential threat. However, surrendering more territory would reinforce the country's loathed position as a "second-class state."[59]

The second additional factor that drove the war's outbreak was Bolivia and Paraguay's historical rivalry over the Chaco Boreal. During negotiations in the late nineteenth century, peaceful dispute resolution seemed possible, as Bolivia focused its territorial claims on the northern Chaco, without challenging Paraguay's authority farther south. The states' early treaties also granted Bolivia port sites above Fuerte Olimpo, satisfying the state's most pressing territorial demand. However, as efforts to ratify the treaties repeatedly failed and the states compiled ever more elaborate documentary evidence to justify their respective territorial claims, positions on the dispute hardened. By 1920, Paraguayans were firmly opposed to any Bolivian port on the Río Paraguay. Meanwhile, Bolivians were articulating ever-greater territorial claims, extending to the confluence of the Río Paraguay and Río Pilcomayo, just outside Asunción. The obvious failure of diplomatic negotiations led many people to conclude that a war over the region was inevitable.[60]

Leaders' increasing obduracy was coupled with rising popular hostility regarding the Chaco dispute. By the end of the nineteenth century, opposition parties in both states had realized that critiquing sitting governments on the Chaco issue was an effective way to foment domestic resentment, thereby advancing their own political agendas. Bolivia's Colorado Party, which dominated national politics from 1888 to 1904, fell because of its inability to favorably resolve the Chaco contest. In Paraguay, President José Patricio Guggiari (1928–1931, 1932) faced a congressional investigation and interruption of his presidency partly because of his perceived mismanagement of the Chaco dispute.[61] School textbooks, newspapers, and the popular press in both states also publicized the contest, intensifying popular engagement. "The result," writes military historian David H. Zook Jr., "was an emotional furor which made compromise difficult."[62] Leaders recognized that any concessions on the Chaco issue would be met with severe domestic political punishment. As a result, they had difficulty backing down in 1927, when clashes provoked a war fervor in both states. After the Laguna Chuquisaca/Pitiantuta confrontations in summer 1932, it was even more difficult to defy popular sentiments.[63]

The final issue that contributed to the Chaco War's outbreak was Bolivian domestic politics. In 1931, when President Salamanca came to power, the state was in the midst of an acute economic crisis. Revenue from tin exports continued to decline, and the government faced major budgetary shortfalls. In July, La Paz announced that it would default on its debts, and that autumn, the state went off the gold standard. Later that year, Salamanca's efforts to increase his executive power and repress unions and communist groups intensified popular hostility toward his regime. Confronted with these seemingly intractable economic and political problems, which had contributed to his predecessor Hernando Siles Reyes's downfall, the president looked for a win somewhere else.[64] He believed that the Chaco could offer one. Paraguay was, in Salamanca's estimation, the one country that Bolivia could defeat.[65] In addition, before becoming president, Salamanca had advocated a more assertive Bolivian position on the dispute.[66] The 1932 clash consequently represented an exceptional opportunity for the president to fulfill his territorial ambitions and rally popular support, thereby enhancing the likelihood of his regime's survival.[67]

The Chaco War was therefore caused by domestic politics, Bolivia and Paraguay's long-standing territorial rivalry, both states' fears of further dismemberment, and Bolivia's desire for an ocean outlet. All of these issues were also intertwined with national pride. Over time, the Chaco contest had become a litmus test for national honor, so both belligerents were defending that, as well as territory, in the Chaco Boreal.[68]

They were not, however, defending or seeking oil resources. La Paz, Asunción, Standard Oil, and informed international observers all recognized that the contested territory's petroleum prospects were poor. The classic oil war explanation emerged as an accusation, an excuse, and a mistake after the conflict began. It was not an accurate interpretation of the belligerents' motives for aggression. It was instead, as Bolivian historian Herbert S. Klein observed, a "mythology . . . universally accepted by all."[69]

The Iran–Iraq War (1980–1988)

The Iran–Iraq War, unlike the Chaco War, was fought in oil-endowed territory. Iraq's invasion, launched on September 22, 1980, concentrated on Khuzestan Province, in western Iran. At the time, Khuzestan produced 80 percent of the state's oil and was home to some of Iran's most important petroleum facilities, including the world's largest oil refinery, at Abadan, and a major oil export terminal at Khorramshahr. Iraqi forces targeted these cities and facilities during their initial advance. This oil-oriented assault, coupled with Iranian leaders' accusations and Iraqi officials' historical claims to Khuzestan, led many observers to identify the conflict as an international petroleum grab. According to this classic oil war interpretation, Iraq attacked Iran in order to seize its neighbor's abundant petroleum resources and revenue.[70] As Andrew Price-Smith states, "desire for the possession of oil fields in the region was the primary cause" of the Iran–Iraq War.[71]

However, this oil war interpretation is mistaken. Although the bulk of Iraqi forces assaulted Iran's richest petroleum province, Baghdad had no intention of holding the territory over the long term. Instead, the Iraqi government aimed to use Khuzestan as a bargaining chip to compel Iran's Islamist regime to accede to its actual war demands: that Tehran cease its destabilizing interference in Iraq's domestic affairs, yield small amounts of contested territory along the central portion of the states' bilateral boundary, and grant Iraq full control over the Shatt al-ʿArab waterway, which forms the southernmost portion of the states' border. Iran's oil resources were not an objective of Iraqi aggression.

In the lead-up to the invasion, Iraqi officials persistently maintained that they had no interest in annexing Khuzestan. In early September 1980, Saddam told his cabinet that he did not "covet" Iranian territory. The same day, Iraq's deputy prime minister, Tariq ʿAziz, announced, "We want neither to destroy Iran nor to occupy it permanently." In a speech to the Iraqi National Assembly on September 17, Saddam reiterated that "Iraq has no ambitions on the Iranian territories." On September 22, the state's Revolutionary Command Council addressed Iranians directly, asserting, "We did not want to harm you; nor do we covet your land."[72]

As their ground offensive began, the Iraqis doubled down on this public messaging. Baghdad Radio broadcast a statement in Persian to anyone within range: "We do not covet Iranian territory." On September 26, Iraqi foreign minister Saʿdun Hammadi claimed that "Iraq has absolutely no designs on Iranian lands" and an Iraqi diplomat reiterated this sentiment to the UN Security Council.[73] Iraqi leaders also explicitly challenged oil war interpretations of their attack. In a news conference on September 24, Iraq's defense minister and deputy commander in chief of the armed forces, Adnan Khairallah, observed that "some foreign news media have begun to hint clearly or between the lines that Iraq wants to control the oil sources in Arabistan [Iraq's name for Khuzestan]." He rejected this argument, reiterating, "We have no designs on either Iranian oil or Iranian land."[74]

These messages, while highly consistent, could have been fabrications; Iraqi officials may have collectively lied about their state's intentions. Iraq's critics, including the Iranian government, were quick to remind international audiences of Baghdad's repeated historical claims to Khuzestan. As they noted, multiple Iraqi governments had asserted that the province should be associated with Iraq, not Iran, because of its large Arab population and historical ties to the Ottoman Empire, Iraq's predecessor state.[75] In international forums like the UN, Iranian officials asserted that Iraq was attempting to seize the province and its oil.[76] Even some Iraqi statements seemed to support this classic oil war interpretation. For example, after the war bogged down in October 1980, Saddam himself threatened

Iran's oil industry, warning that a prolonged Iraqi presence in Iranian territory "creates certain rights which did not exist before the war began."[77]

Yet there are numerous reasons to question classic oil war interpretations of Iraq's invasion. First, Iraqi authorities repeatedly, publicly promised to withdraw from Khuzestan after achieving their war aims. On the first day of the invasion, Iraq's Revolutionary Command Council stated, "We shall withdraw from the Iranian territory in which our presence is necessitated by our defensive military requirements and army's security as soon as Iran recognizes our sovereign territory and respects our vital interests."[78] In press conferences on September 25 and 26, ʿAziz stressed that Iraq would "withdraw from the Iranian territories it is occupying" if Iran agreed to recognize Iraq's sovereignty over the Shatt al-ʿArab and contested central border territories, cease its interference in Iraq's domestic affairs, adhere to "good-neighbor relations," and withdraw from occupied islands in the Persian Gulf.[79] Three days later, Saddam offered a ceasefire and negotiations if Iran would accede to those demands. At a press conference in November, the Iraqi president repeated the offer, asserting, "Any time the Iranian officials . . . recognize our rights, then we will withdraw from their land."[80]

At the same press conference, Saddam explained that his threats to Iranian oil resources had been tactical rather than an expression of his state's actual war aims. By invading Iran, he asserted, Iraq was "twisting the Iranian rulers' arms. If we find that a certain degree of twisting is not enough, we will add another degree to it."[81] In October, Iraq's deputy prime minister, Taha Yasin Ramadan, had issued a similarly conditional threat. He asserted that "Arabistan's oil will be Iraqi as long as Tehran refuses to negotiate," but also pledged that Baghdad would control the resources only "until a solution is found."[82] The public nature of these pronouncements raised the cost of reneging on them. Any Iraqi backpedaling, including a sustained occupation of Iranian territory and oil installations, would incur international costs. Not only would the regime have violated the norm against conquest, it would also have lied to the international community, inviting further censure.

Hence, another reason to doubt classic oil war interpretations of the Iran–Iraq conflict is the Iraqi leadership's demonstrable awareness of the obstacles to fighting for oil. In addition to recognizing the international impediments to conquering Khuzestan, they were cognizant of invasion obstacles to seizing oil-endowed territories. In 1973 and 1974, Iraq and Iran refrained from broadening a series of bilateral border clashes partly because both governments feared that an outright war would damage their petroleum industries. As Saddam noted in February 1975, "[Oil] is a very inflammable material."[83] Before attacking Khuzestan in 1980, Iraq's leader acknowledged that a major conflict with Iran could threaten his own state's petroleum infrastructure.[84] Saddam decided to take that risk, nonetheless, because he was pursuing other goals, not oil.

Iraqi officials were also sensitive to the occupation obstacles to seizing Khuzestan. They knew that controlling the province over the long term was unviable. Although they believed that Khuzestan's residents would accede to a temporary Iraqi military presence because of their shared Arab ethnicity and the population's long-standing desire for greater autonomy from Tehran, they recognized that the local population sought self-government within Iran, not secession.[85] As ʿAziz noted in a May 1980 interview, the people of Khuzestan aspired to "autonomy within the framework of a single Iranian state."[86] Locals would therefore resist a prolonged Iraqi occupation and stridently oppose their province's forced annexation to Iraq. Hence, while Baghdad persistently advocated self-government for Khuzestan, it did not seek to conquer the territory. Instead, Iraq's claim was "largely rhetorical."[87]

Iraqi officials also had little incentive to prosecute a classic oil war because, had they wished to boost national petroleum revenue, they possessed a far less costly means of doing so: expanding domestic oil output. Iraq's petroleum rents had already risen enormously in the decade before the war. After fully nationalizing the oil industry in 1972, the state rapidly intensified exploitation of its historically underdeveloped reserves. In addition, the first and second energy crises (1973–1974 and 1978–1979) triggered sharp jumps in international oil prices. These two developments increased Iraq's petroleum revenue from $575 million in 1972 to an astonishing $26 billion eight years later.[88] On the eve of the war, there was still scope for further expansion. Accordingly, if the government wanted to augment its already abundant oil income, it could intensify resource exploration and development, rather than invading its neighbor. Iraq's defense minister, Khairallah, highlighted this point when he rejected classic oil war interpretations of the conflict. "You know very well that we are not in need of the Arabistan oil," he told reporters. "We have enough oil."[89] War historian Edgar O'Ballance concurs, observing that Iraq had "ample reserves" of petroleum, so oil ambitions were not "one of the deep-rooted catalysts" of the war.[90]

Iraqi leaders, including Saddam, were therefore aware of the invasion, occupation, and international obstacles to conquering Khuzestan. They also lacked the impetus to confront these impediments, since they possessed an alternative, less costly means of increasing national resource revenue. It is consequently unsurprising that the Iraqis pledged to withdraw from the province as soon as they achieved their war aims. They were not trying to grab Iran's oil, so there was no need to stay.

Rather than fighting for oil, Iraq's central reasons for attacking Iran were threefold. One was historical rivalry; the states have competed for dominance in the Persian Gulf region for more than a millennium, and this contestation intensified in the two decades before the war. The second motive was territorial; although

Iraqi leaders did not aim to conquer Khuzestan, they did aspire to regain full authority over the Shatt al-ʿArab waterway and some "usurped" territories along the central portion of the international land boundary.[91] The third, and most important, reason for Iraq's attack was national security; Saddam aimed to weaken Iran's revolutionary Islamist government, which he perceived as an increasing threat to his regime's survival.[92]

The first incentive for aggression, Iran and Iraq's historical rivalry, dates back to at least the seventh century, when Arabs conquered the Sasanian Persian army in the Battle of al-Qadisiyyah. In the mid-twentieth century, the modern states remained two of the Middle East's most powerful countries and were still divided by ethnicity and religion; the majority of Iraqis are Arabs and traditionally ruled by Sunnis, while most Iranians are Persian and Shiʿa. These ascriptive divides contributed to contemporary animosity. However, between the late 1950s and mid-1970s, two additional political developments intensified the states' bilateral hostility.[93] One of these was the Iraqi revolution (1958), which overthrew the state's Hashemite monarchy. Iraq's new Republican regime was more aggressively nationalistic than its predecessor and more assertive in its territorial claims. In late 1959, Iraqi prime minister Abd al-Karim Qasim pointedly highlighted Khuzestan's historical associations with the Ottoman Empire and proclaimed Iraq's authority over the entire Shatt al-ʿArab waterway.[94]

The second contemporary development was the United Kingdom's withdrawal from the Persian Gulf in December 1971. This retreat created a power vacuum in the region, which British and American authorities encouraged Iran to fill, as they believed that Shah Mohammad Reza Pahlavi was the best local substitute for Western authority.[95] The United States and Britain provided the shah with arms and diplomatic support to strengthen his regional position, antagonizing Iraq's new Baʿathist regime, which had seized power in July 1968 and adopted a more overtly anti-Western stance than its predecessor.

Over the next decade, both of the rival states employed a variety of strategies to shift the regional power balance in their favor. One was direct attacks; during the spring of 1972 and winter of 1973–1974, Iraq and Iran engaged in repeated clashes along their shared boundary. A second strategy was territorial expansion. In November 1971, on the eve of Britain's withdrawal from the Gulf, Iran seized several islands near the Strait of Hormuz, which had belonged to Sharjah and Ras al-Khaimah, two of the future United Arab Emirates. Many Arab states, including Iraq, protested the confiscation.[96] However, Iran's growing regional clout and great power backing left them powerless to reverse it.

Third, Baghdad and Tehran supported opposition movements within each other's territory. The Iranians provided significant military and financial support to Iraq's Kurds, who were waging an armed struggle for autonomy against Baghdad. The Iraqis supplied aid to Arab autonomy movements in Khuzestan and antigovernment groups in Iranian Kurdistan and Balochistan. These Iraqi initiatives had little impact on Iran.[97] However, by the mid-1970s, Tehran's support to Iraq's Kurds constituted a major security threat to Baghdad. The Kirkuk oil fields, which contain up to 40 percent of Iraq's oil reserves, are located in Iraqi Kurdistan. Consequently, the state would lose a significant portion of its petroleum revenue if the Kurds obtained independence. By late 1974, Baghdad was on the brink of losing control of the territory. Although the government had deployed one hundred thousand troops to the region, it was unable to contain a renewed Kurdish uprising, bolstered by Iranian aid.[98]

To curtail Tehran's support for the Kurds, the Baʿathists were compelled to participate in the Algiers Agreement (1975), in which Iran and Iraq pledged to cease their interference in each other's domestic affairs. The accord also contained provisions for the states to shift their boundary in the Shatt al-ʿArab from the river's eastern bank to the *thalweg* (primary navigable channel) and delimit their land boundary in accordance with the Constantinople Protocol (1913) and the Proceedings of the Border Delimitation Commission (1914). The Iraqis perceived the former provision as a major concession and only acceded to the Algiers Agreement because they believed that they had no other choice. As Foreign Minister Hammadi stated, "It was either that or lose the country."[99] Saddam, who signed the accord as Iraq's vice president, described it as "his only political failure." He felt personally humiliated by the agreement and was determined to reverse it.[100]

Iraq's territorial goals—the state's second motive for the Iran–Iraq War—were directly related to the Algiers Agreement. Baghdad aspired to return the Shatt al-ʿArab boundary to the river's eastern bank and regain control over small amounts of territory located along the central portion of the states' boundary, which Iraqis believed had been promised to them in the accord. The first ambition initially appears dubious; why was Iraq so invested in control over half of a waterway? Yet the riparian dispute had been one of the central sources of bilateral animosity since the mid-nineteenth century. In 1847, the second Treaty of Erzurum recognized Iraq's control over the entire river, setting the states' boundary along the waterway's eastern bank, rather than the thalweg. This unusual delimitation method aimed to compensate Iraq for its diminutive Persian Gulf coastline, which is only forty miles long.[101] Iraq was therefore highly reliant on its Shatt al-ʿArab ports, whereas Iran possessed expansive coastal territories. The 1847 treaty nonetheless granted Iranians the right to navigate the waterway and recognized Tehran's authority over the port of Khorramshahr and the island of Abadan.

Between 1847 and 1975, Iraq's control over the waterway gradually diminished. In the Constantinople Protocol, Iran gained authority over the anchorage at Khorramshahr, and during the subsequent delimitation process, the riparian boundary

MAP 5.2. The Iran–Iraq border (1980)

around the city was shifted to the thalweg to accommodate the port's increasing traffic, driven by Iran's burgeoning oil industry.[102] In 1937, the Tehran Treaty moved the international boundary to the thalweg for several additional miles around Abadan. During the 1940s and 1950s, the Iraqis successfully evaded repeated Iranian requests for joint administration of the waterway. They also resisted Tehran's demand, in 1961, that all ships entering Iranian ports be piloted by Iranians. The Basra Port Authority, the Iraqi organization that handled river traffic, responded to this initiative by barring passage to all Iranian ports along the waterway. Oil exports from Abadan plunged 60 percent and, after nine weeks of crippling blockade, Tehran was forced to rescind the order.[103]

However, when power in the Persian Gulf shifted in the late 1960s, so did authority in the Shatt al-ʿArab. On April 19, 1969, the shah abrogated the Tehran Treaty and claimed a thalweg boundary for the entire waterway. To underscore this change, Iranian commercial vessels, piloted by Iranians and escorted by the Iranian navy, traveled up the river without paying navigational tolls to the Basra Port Authority. The Iraqi regime, beset by instability in Kurdistan, was unable to respond to the challenge directly. Instead, it sent letters of protest to the UN Security Council, increased funding for Khuzestan separatists, and expelled thousands of Iranians from Iraq.[104] Yet these actions failed to alter Iran's river behavior. From that point on, the Shatt al-ʿArab's de facto boundary was the thalweg; the Algiers Agreement merely formalized the riparian status quo. After the accord, the Iraqi government was committed to shifting the boundary back to the eastern shore, both to secure control over a vital transportation route and to restore national pride.[105]

Iraq's second territorial goal in the Iran–Iraq War was to regain sovereignty over approximately 130 square miles of territory located along the central portion of the states' land boundary, between Qasr-e Shirin and Mehran. The Iraqis believed that the Algiers Agreement compelled Iran to return those territories, which the state had gradually occupied between 1913 and 1975.[106] The areas were strategically located along the main transit route from Tehran to Baghdad and included higher elevations, at the foot of the Zagros Mountains, overlooking Iraq's Diyala Province.[107] Iraq's broader interests in the territories, like its interests in the Shatt al-ʿArab, were therefore to strengthen state security and restore national honor.

The Baʿathist government initially attempted to regain the border territories peacefully. The states began to delimit their land boundary, as stipulated by the Algiers Agreement, in May 1978.[108] However, before the border issue was fully resolved, the Iranian Revolution interrupted the process. On January 16, 1979, the shah fled Iran, bringing Ayatollah Ruhollah Khomeini to power. Acknowledging the disruption, the Iraqis accepted a temporary pause in delimitation activity. However, as time passed, they became increasingly convinced that the new Iranian regime had no intention of ever completing the procedure. As Hammadi stated on September 10, 1980, "When the revolution took place in Iran, the Iraqi Government gave the new regime some time to organize its affairs, so that it would be in a position to complete the handing over process. But the abovementioned territory was not delivered to the Iraqi side. On the contrary, as time passed the Iranian officials showed hostile intentions and the inclination to expand at the expense of Iraqi and Arab territory."[109] Iran's president, Abolhassan Banisadr, eventually confirmed Iraqi suspicions; his regime did not plan to fully implement the accord.[110]

Saddam concluded that the only way to retake the central border territories and regain control over the Shatt al-ʿArab was by force. On September 7, 1980, Iraqi troops began to seize the contested areas, beginning around Zain al-Qaws. Over the next few days, they occupied additional territory around Saif Saad and Qasr-e Shirin.[111] In a meeting with his advisers on September 16, six days before the Iran–Iraq War's internationally recognized start date, Saddam asserted that Iraq had successfully retaken all of the territories promised to it in the Algiers Agreement. As he put it, "Today we can say that all of the lands extorted by Iran are back under our sovereignty."[112]

However, the Shatt al-ʿArab issue remained unresolved. Saddam and his advisers recognized that retaking the waterway would be more contentious than seizing the central border areas and could escalate into a "full-scale war."[113] Nonetheless, they were determined to proceed. On September 17, Saddam publicly abrogated the Algiers Agreement, tearing up his copy of the accord on state television. The Iraqi government also attempted to reassert control over river navigation by compelling foreign ships to raise the Iraqi flag and accept guidance by the Iraqi navy.[114] Fighting in the Shatt al-ʿArab, which had been limited up to that point, intensified; Iran deployed its navy and began firing on Iraqi patrol boats.[115]

On September 23, Iraq launched a major ground offensive in several of Iran's western provinces. This assault aimed to secure Iraq's control over the contested central border territories and the Shatt al-ʿArab. However, its primary goal—and Iraq's third and most important reason for prosecuting the war—was to moderate, or ideally eliminate, the Islamist regime's threat to the Iraqi government. This menace was unanticipated. Iraqi leaders had initially welcomed the Iranian Revolution, as they were pleased to see the shah replaced by a movement that was hostile to Western intervention in the Gulf region. In February 1979, Saddam had reached out to Iran's new government, stating that "a regime which does not support the enemy against us and does not intervene in our affairs, and whose world policy corresponds to the interests of the Iranian and Iraqi people, will certainly receive our respect and appreciation."[116] However, by that summer, the prospect of improving bilateral relations seemed remote.

Ideologically, Iran's Islamist regime was diametrically opposed to Iraq's Baʿathist government. The Baʿath Party was secular, nationalistic, and committed to pan-Arabism, which aspired to informally unite the Middle East under the banner of shared ethnic identity. The Islamists were Persian, intensely religious, and aspired to build a universalist, pan-Islamic republic.[117] More worrisome than these ideological differences, however, were Iranian leaders' public attacks on the Iraqi regime. In spring 1979, Iraq experienced a resurgence of Shiʿa activism, including protests in the cities of Najaf and Karbala.[118] The Baʿathists attempted to contain the movement by declaring membership in the Daʿwa Party, a Shiʿa organization founded in the late 1960s, a capital offense. However, in June 1979, Khomeini and his followers began to encourage Iraqis to overthrow their "infidel" leaders. By the end of the year, the Iranian regime was actively supporting al-Daʿwa and had resumed its assistance to Iraq's Kurds. In December, Iraqi intelligence sources reported that Iran, Syria, and Libya were collaborating to provoke a sectarian civil war in Iraq, which would overthrow Saddam's regime.[119]

Relations deteriorated further in spring 1980, as Iranian officials intensified their calls for the Iraqi regime's overthrow. On March 15, Khomeini urged Iraqi citizens to revolt against their government, and less than a week later, his son called for the "export" of the Iranian Revolution. On April 8, Khomeini said that Iraqis should "wake up and topple this corrupt regime in your Islamic country before it is too late." Nine days later, he inveighed that "Iran will break Iraq and advance to Baghdad."[120] Earlier that month, Iran had supported an assassination attack against Tariq ʿAziz, which was claimed by al-Daʿwa.[121] In June, Foreign Minister Sadegh Ghotbzadeh "revealed . . . that his government had taken the decision to topple the Baath regime."[122]

Escalating border skirmishes intensified the Iraqi government's sense of peril. On April 4, 1980, Banisadr put Iranian border forces on alert and warned that Iran would go to war if the situation deteriorated further. Clashes subsequently escalated, and in May, internal Iraqi memoranda from the air force and Air Defense Command asserted that, over the previous forty-five days, Iran had committed thirty-seven border violations. Many of Saddam's generals, interviewed after the 2003 US invasion, stated that they were anticipating an Iranian attempt to overthrow their government, either through direct attacks or via support for Iraq's domestic opposition groups.[123]

By summer 1980, Saddam had concluded that Iranian interference constituted an existential threat to his regime.[124] He also believed that his options for resisting the threat were increasingly limited. A failed military coup in Tehran on July 9–10 convinced Saddam that Iran's internal opposition would not be able to topple the Islamists.[125] Hence, Iraq would have to intervene directly to eliminate the threat. In mid-July, the president met with his chiefs of staff and instructed them to prepare for war with their neighbor. On August 16, Saddam told his generals that Iraq would definitely attack, although he left the date unspecified.[126]

Meanwhile, border skirmishes continued to intensify. By early September, the Iraqi National News Agency was reporting persistent Iranian attacks, many of them launched from the contested central border zones. According to Iraqis, the Iran–Iraq War began on September 4, 1980, when the Iranians purportedly launched intensive artillery assaults on Iraqi border posts at Khanaqin, Mandali, Badra, and Zurbatiyah.[127]

Iraq's goals, in responding to these assaults and, eventually, invading Iran, were to eliminate a perceived existential threat to the Baʿathist regime, regain control over lost territories, and shift the regional power balance in Iraq's favor. The state did not aim to annex Khuzestan and seize its oil resources. The Iraqi invasion nonetheless targeted Khuzestan for several reasons, some of them linked to the province's petroleum endowments. First, the territory is relatively accessible. In comparison with the central border area, where Iranian territory climbs rapidly into the Zagros Mountains, Khuzestan is fairly flat, which permitted a longer military advance. Second, the Iraqis believed that Khuzestan's Arab population would offer little resistance to their invasion. Third, attacking Khuzestan would interrupt Iranian oil production and transportation, severing the regime's vital petroleum supplies and income.[128] All of these factors gave an Iraqi attack in Khuzestan greater impact than an attack elsewhere, increasing the likelihood that the Iranian government would accede to Iraq's demands.

Iraq's strategic aims in attacking its neighbor were nonetheless imprecise.[129] However, an invasion of Khuzestan could moderate the perceived Iranian menace in at least four ways. First, a forceful demonstration of Iraq's military power, coupled with Iranian battlefield losses, might push the Islamists to yield to Baghdad's demands to cease their interference in Iraqi domestic affairs and cede control over the contested territories. Second, an invasion could draw military forces away from Tehran, leaving Khomeini's regime exposed and vulnerable to a domestic coup.[130] Third, attacks on Khuzestan's oil infrastructure could facilitate the Islamists' overthrow by "paralyzing the Iranian economy" and undermining national defense.[131] Fourth, by supporting Khuzestan's long-standing bid for self-government, Iraq might detach the province's oil from Tehran, permanently reducing the amount of petroleum revenue that the Iranian government could use to foment unrest in Iraq.

Khuzestan's oil therefore contributed to Iraqi leaders' decisions about how to prosecute their war with Iran. However, the province's petroleum resources were not a prize that the Baʿathists hoped to win. Instead, they were prepared to withdraw from the region after accomplishing their actual war aims. The classic oil war interpretation of Iraq's aggression was a wartime construction, often asserted with little supporting evidence.[132] It gained additional traction after Iraq invaded Kuwait in 1990, as, in the wake of that oil gambit, observers were quick to assume that Saddam's earlier aggression had also targeted petroleum.[133] Subsequent authors have repeated this interpretation; as Pierre Razoux surmised in his recent history of the Iran–Iraq War, "One could also conceive that Saddam's thuggish disposition instinctively led him to attempt to rob Iran's oil reserves."[134] Yet, as this chapter's two case studies have demonstrated, intuitive assumptions often fail to hold up to critical scrutiny. The Iran–Iraq and Chaco conflicts were red herrings, not classic oil wars.

6

OIL SPATS

The Falkland/Malvinas Islands Dispute

The primary difference between red herrings and oil spats, the second type of conflict involving hydrocarbon-endowed territories, is that the latter are motivated primarily by states' petroleum aspirations. In oil spats, aggressors aim to obtain control over petroleum resources or defend their resource claims against the initiatives of other oil-seeking states. Usually, oil spats are precipitated by unilateral resource exploration or development. One country attempts to exploit contested oil or natural gas reservoirs, and another country forcefully intervenes to interrupt them. These confrontations often provoke hostile rhetoric while they are under way. However, the participants' military actions are universally mild. Aggressors deploy a warship, mobilize a small number of soldiers, engage in minor border violations, attempt to intercept each other's ships or drilling rigs, or cut the cables of seismic survey vessels. They do not engage in major clashes. Moreover, no oil spat has ever spiraled into an oil campaign or oil gambit; those larger conflicts are provoked by other issues and interests, rather than emerging from petroleum sparring. Contrary to popular perceptions, oil spats are a minor threat to international security.

This chapter presents a representative case study to illustrate oil spats' dynamics, including their failure to escalate: Argentina and the United Kingdom's Falkland/Malvinas Islands dispute. London and Buenos Aires have competed for sovereignty over the contested islands since the 1830s and, in the course of their disagreement, have prosecuted one oil spat and one severe interstate conflict. The latter was the Falklands War (1982), which some authors have linked to petroleum ambitions.[1] In addition, since 2010, renewed efforts to develop the islands' oil have prompted assertions that petroleum aspirations could inspire further Anglo–Argentine violence.[2] However, as this chapter will demonstrate, these predictions drastically overstate the states' interest in fighting for oil resources.

Argentina and the United Kingdom's one oil spat occurred in February 1976, when an Argentine destroyer approached a British research ship, the RRS *Shackleton*, sailing near the contested islands. Argentine authorities believed that the ship was connected to a British scientific survey, which they assumed was searching for oil in the Falklands' continental shelf. To discourage this unilateral initiative and defend Argentina's resource claims, the destroyer attempted to intercept the *Shackleton*, firing several warning shots across its bow. The confrontation provoked diplomatic protests from both states. However, it did not intensify; neither participant initiated further military operations. In addition, while oil interests drove the *Shackleton* spat, they were not Argentine officials' only concern; leaders also aspired to reassert their state's claim to the disputed islands, pressure the British government to accelerate sovereignty negotiations, and rally domestic political support.

The Falklands War, Britain and Argentina's severe military conflict, was not motivated by petroleum ambitions; instead, it was another red herring. By the early 1980s, both states' interest in the islands' oil had waned. In addition, had Argentina wanted to boost its revenue from the Falklands' resources, it could have done so without a war. Nor did lingering animosity from the 1976 oil spat encourage contention. Instead, the war's fundamental cause was Argentina's long-standing desire to regain island sovereignty. Its triggers were the impending sesquicentennial of Britain's occupation, the Galtieri government's concern about its political survival, the evident failure of diplomatic negotiations to resolve the prolonged dispute, anticipation of an imminent shift in the regional balance of power, and Argentine miscalculation of the consequences of the attack.

The *Shackleton* Spat (1976)

The Falkland/Malvinas Islands are a barren archipelago in the South Atlantic Ocean, 300 miles from mainland Argentina and 7,500 miles from the British Isles. The islands' population has never exceeded three thousand, and the territories have struggled economically as a result of their geographic isolation and unforgiving natural environment. Nonetheless, for over 180 years, the Falklands have been the object of a bitter sovereignty dispute between the United Kingdom and Argentina. The contest began in 1833, when British forces occupied the islands, pushing out a small contingent of Argentine soldiers. Buenos Aires protested, but the British refused to withdraw.[3]

For more than a century, interest in the dispute was limited in both states. However, Argentine engagement mounted after World War II, when President Juan Perón discovered that he could use the contest to foster nationalist enthusiasm and popular support. Perón's successors also played up the islands issue, initiating a Malvinas Day and Malvinas museum in 1963.[4] British engagement with the dispute also increased in the 1960s, as the global decolonization movement gained momentum. The British Colonial Office was attempting to relinquish many of its overseas possessions in order to mollify international public opinion and reduce national expenditures. To advance these goals, the state was prepared to grant self-determination to the Falkland islanders.

However, Argentine authorities resisted this shift. Conscious of the islanders' strong sense of attachment to Britain, they feared that, if allowed to decide on their own political future, the Falklanders would choose continued "free association" with their former metropole rather than integration into Argentina. Buenos Aires therefore persuaded the UN to identify the Falklands as a contested territory. On December 16, 1965, Resolution 2065 recognized the existence of a dispute between Argentina and the United Kingdom and encouraged the countries to resolve it peacefully, taking into consideration the interests of the islanders.[5]

Initial negotiations, conducted from 1966 to 1968, failed to produce a jointly acceptable settlement. The British Foreign Office would have happily transferred authority over the Falklands to Argentina.[6] The islands' economy was floundering, defending them was costly, and the dispute was harming Britain's broader relations with Argentina, an important trading partner.[7] However, British officials also believed that they were morally bound to consider the islanders' political preferences.[8] The Falklanders' position was firm; they were determined to remain attached to the United Kingdom. They also mustered considerable support for their position in Parliament. In 1968, members of Parliament repeatedly excoriated the Foreign Office's attempt to "sell out" the Falklands. For their part, Argentine authorities were unwilling to accept anything short of full island sovereignty. The parties' positions were therefore irreconcilable.

In the midst of this diplomatic impasse, oil prospects entered the picture. By 1969, a subsidiary of Royal Dutch Shell had begun exploring for petroleum in uncontested portions of Argentina's continental shelf, off Tierra del Fuego. Some companies anticipated that these potentially oil-bearing sediments might extend to the Falkland Islands. Consequently, several of them approached the British and Falkland Islands governments to express their interest in exploring the islands' continental shelf. However, when the British cabinet discussed the issue in October 1969, the Foreign Office resisted further action. Officials preferred to conceal information about the Falklands' prospective petroleum resources and avoid exploration, out of fear that unilateral activities would provoke Argentine hostility.

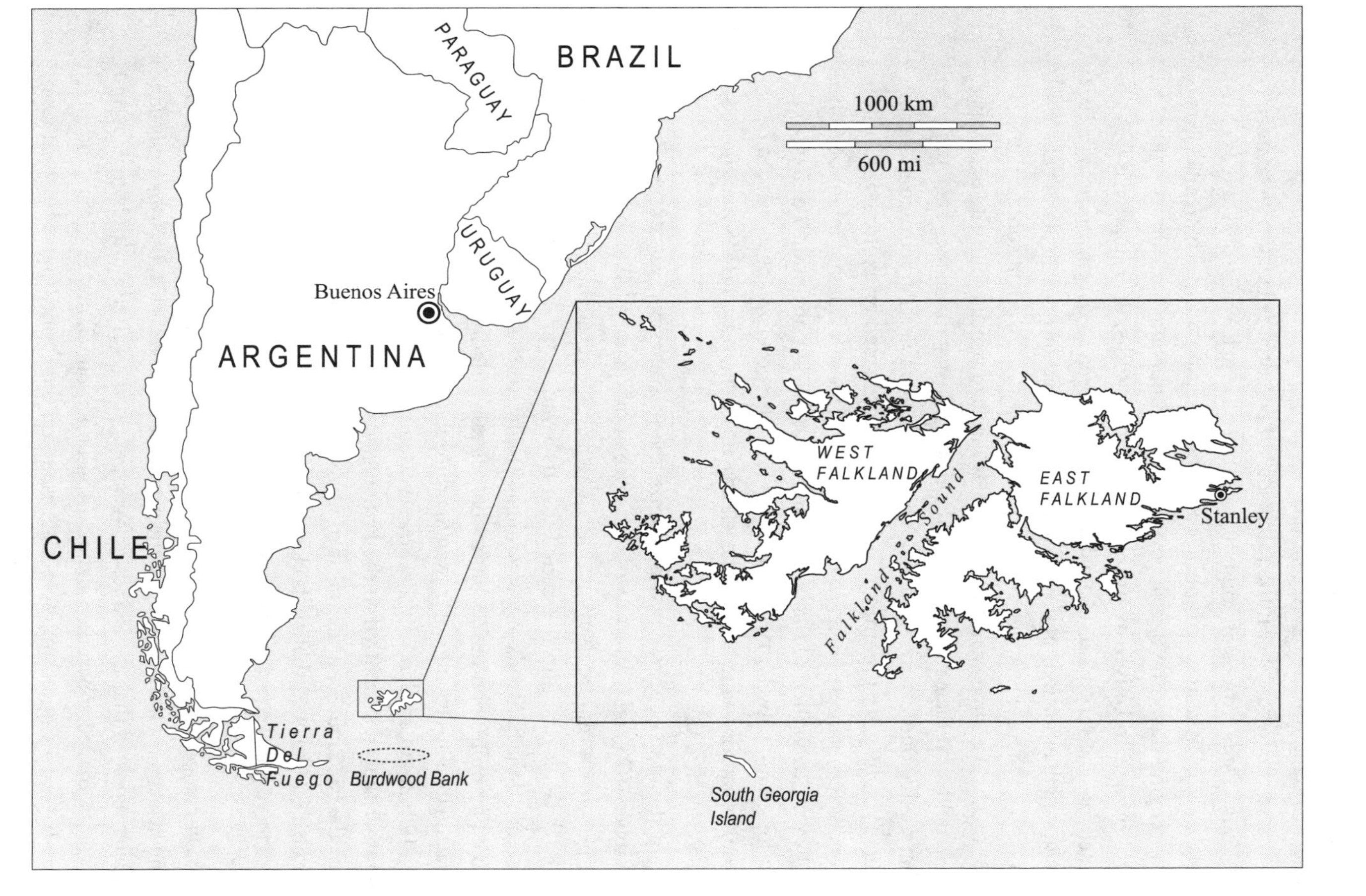

MAP 6.1. The Falkland/Malvinas Islands (1976)

If any exploration was to take place, they believed, it would have to be collaborative, involving Britain and Argentina.[9]

These efforts to conceal the Falklands' oil potential failed. On November 27, 1969, the *Daily Express* of London published an article entitled "There Is Oil off Falklands."[10] In response, inquiries about exploration licenses increased. To justify denying them, the British government commissioned a "library study," led by Birmingham University professor Donald Griffiths, to evaluate the area's oil potential. The survey's results, which were conveyed to the government in 1971, suggested that portions of the Falklands' continental shelf, particularly near Burdwood Bank, approximately 120 miles south of the islands, might contain petroleum resources. The same year, the companies that had processed Shell's exploratory data published their results in *World Oil*, an industry magazine. They suggested that the region near the islands might contain commercial petroleum deposits. These results were preliminary and not yet confirmed by exploratory drilling. Nonetheless, rumors began to circulate that the South Atlantic might be a "new Kuwait." To gather additional information—and continue to delay private exploration—between 1971 and 1974, the British government sponsored further geophysical surveying around the islands, using the vessels HMS *Endurance* and RRS *Shackleton*.[11]

In the midst of these investigations, the first energy crisis struck. Between October 1973 and January 1974, the price of oil quadrupled. Suddenly, interest in petroleum resources around the Falklands skyrocketed. A 1974 report from the US Geological Survey asserted that the continental shelf between Tierra del Fuego and the Falklands might contain 40 billion to 200 billion barrels of oil. That would be more than the 35 billion barrels purportedly contained in the North Sea and perhaps even greater than Saudi Arabia's reserves, estimated at the time at 119 billion barrels. Griffiths, however, tempered these optimistic findings. In another report, submitted to the British government in 1975, he acknowledged that the Falklands' continental shelf might contain commercial oil deposits. But he also emphasized that such claims were still speculative.[12] Other commentators highlighted the challenges of operating around the Falklands, including high seas, dramatic tidal shifts, and deep water.[13]

Oil companies were nonetheless intensifying their efforts to obtain exploration licenses for the Falklands' continental shelf. In late 1974, the *Falkland Island Times* reported that the islands' government had received multiple new permit requests. In December, the Falklands' legislative council met to discuss the oil issue. Eager to exploit any potential resources, the council passed two motions asking the governor of the islands, a British official, to invite applications for exploration licenses. Although the governor did not support the initiatives, oil companies continued to submit inquiries.[14]

These applications were not firm commitments to search for oil. Companies were eager to demonstrate their interest in the Falklands and obtain exploration rights in order to stymie potential competitors. Yet actual exploration was widely perceived as unfeasible, because of the strong likelihood of Argentine opposition.[15] As Robin Edmonds, an assistant undersecretary at the Foreign Office, observed, "Clearly, no oil company would spend a single dollar looking for anything in an environment as hostile as that around the Falklands, unless they had a reasonable certainty that there was not going to be a tremendous intergovernmental row about the outcome."[16] Companies also did not want to antagonize Argentina, as that could imperil their ability to participate in other South American oil projects. Nor did they wish to expose themselves to the risks of investing in a disputed territory, particularly one that was difficult to reach and supply. The British government was unwilling to provide naval protection to deter Argentina from interfering with petroleum projects.[17] Hence, the investment obstacles associated with developing the area's petroleum resources were high, even in the absence of outright military confrontations.

Argentine officials had made their opposition to unilateral British oil exploration abundantly clear. In December 1974, Argentina's representative to the UN told the General Assembly that all subsurface resources around the Falklands belonged to Argentina and that no effort should be made to exploit them until the sovereignty dispute was resolved. The Argentine foreign minister, Alberto Vignes, reiterated this position the next spring, stating that Buenos Aires did not recognize a foreign government's rights to explore for or extract oil resources around the islands. Officials also asserted that any unilateral exploration would be a breach of UN resolutions.[18] Noting the increasing, resource-related tensions, London's *Sunday Telegraph* asserted on March 9, 1975, that oil competition could provoke Anglo–Argentine military clashes.[19]

These fears proved prescient. Still seeking to facilitate an eventual transfer of sovereignty to Argentina, in summer 1975, the Foreign Office commissioned a survey to evaluate the Falklands' long-term economic viability. The bureau recruited Lord Edward Shackleton, son of famed Antarctic explorer Ernest Shackleton, to lead the expedition. Before the survey's departure, Foreign Secretary James Callaghan explicitly informed Lord Shackleton of what he was expected to find: that the islands had "no economic future" without closer linkages with Argentina.[20] However, the British failed to adequately prepare Buenos Aires for the expedition. Callaghan mentioned the survey to Argentina's foreign minister, Angel Robledo, in September 1975. Yet, poor translation caused Robledo to conflate the survey with another of Callaghan's proposals, for a senior British official to travel to Buenos Aires for talks on economic cooperation. As a result of this miscommunication, Argentine officials did not become aware of the

Shackleton expedition until mid-October, when the British government publicly announced it.[21]

Argentine authorities were aggravated by the apparent lack of prior consultation, and, rather than cooperating fully with the survey, as the Foreign Office had expected, the increasingly unstable Perón regime—now headed by the former ruler's third wife, Isabel Martínez de Perón—decided to bolster its domestic support by challenging Britain's initiative, which was intensely unpopular within Argentina.[22] The Ministry of Foreign Affairs reiterated that all natural resources around the Falklands belonged to Argentina, claimed that the British survey "violates the [UN] principle of abstaining from new unilateral initiatives," and refused permission for the survey team to transit Argentine territory. In November and December, civilian and military officials asserted that British research ships must seek Argentina's permission to operate within the country's two-hundred-nautical-mile continental shelf—including the areas around the islands—and warned that, if they failed to do so, they would be taken into custody. Attempts to mollify Buenos Aires by including Argentine scientists in the survey failed. As popular hostility toward the expedition mounted, the Péron government feared a domestic backlash for any cooperation with the Shackleton research team.[23]

The survey proceeded in spite of this opposition. Shackleton's team sailed from southern Brazil to the Falklands on the HMS *Endurance* in late December. They then committed another diplomatic blunder: landing on the islands on January 3, 1976, the anniversary of Britain's occupation. Argentine hostility toward the survey intensified; Foreign Minister Manuel Guillermo Arauz Castex described the timing as "an unfriendly and unhelpful coincidence."[24] Soon after, Castex requested that London withdraw its ambassador from Buenos Aires. He also instructed Argentina's ambassador to the United Kingdom, who was on leave, not to return to London—effectively, although not formally, severing the states' diplomatic relations.[25]

The countries' oil spat occurred on February 4. An Argentine destroyer, the ARA *Almirante Storni*, intercepted the RRS *Shackleton* seventy-eight miles south of the Falklands' Cape Pembroke.[26] The *Shackleton* was unconnected to the Shackleton survey; instead, it was reportedly investigating continental drift in the South Atlantic. However, the *Storni*'s captain mistook the ship for the *Endurance*, which had carried the research team to the islands.[27] After hailing the *Shackleton* as *Endurance*, the *Storni* demanded that it stop its engines and allow Argentine sailors to board. When the *Shackleton*'s captain, Philip Warne, refused, the *Storni* fired several warning shots across his bow. With an Argentine naval aircraft flying overhead, the warship also threatened to fire on *Shackleton*'s hull but refrained after Warne broadcast that his ship was carrying explosives for geophysical research. Warne also ignored the *Storni*'s demand to proceed to the Argentine port of Ushuaia; instead, the captain "turned a Nelsonian blind eye" and continued to his original destination, Port Stanley. The *Storni* pursued the *Shackleton* until the vessel was within six miles of the islands.[28]

The confrontation was driven primarily by oil. Since the Shackleton survey was announced, Argentine officials had suspected that its central aim was to investigate the Falklands' petroleum prospects.[29] Argentina's ambassador to the UN, Carlos Ortiz de Rozas, stated as much when he denounced the survey before the General Assembly in December.[30] When the expedition reached South America later that month, Argentine nationalists protested and newspapers accused it of searching for oil.[31] Describing the fervor, an *Economist* article of January 24, 1976, reported that, when the Shackleton survey arrived, "suspicious Argentine minds concluded that the British were after the islands' only other likely asset: oil."[32] After the spat, Ortiz de Rozas claimed that the *Shackleton* had also been searching for petroleum.[33]

The last accusation was plausible, if inaccurate; as discussed earlier, both the *Shackleton* and the *Endurance*, which transported Shackleton's research team to the islands, had conducted seismic surveys in the region earlier that decade. To combat these widespread, oil-related beliefs, the Foreign Office dispatched Minister of State Edward Rowlands to meetings in New York the next month, armed with documentary evidence that the *Shackleton* had not been exploring for oil.[34] The British government also tried to tamp down hostility toward the Shackleton survey by reiterating that it aimed to evaluate the islands' broader economic prospects, rather than its petroleum potential specifically. These efforts nonetheless failed to dislodge popular Argentine assumptions that Britain was maintaining its hold over the islands for their prospective petroleum resources.[35]

The *Shackleton* incident is a highly representative oil spat. It revolved around the development of contested oil resources and entailed one state attempting to interrupt another's unilateral resource exploitation. The aggressor's goal was less to grab petroleum reservoirs than to prevent another claimant from monopolizing control over them. Like most oil spats, the incident occurred in disputed territory, between two states that shared a history of hostility. Participants were therefore inclined to believe the worst of each other, which encouraged them to respond to even minor, petroleum-related provocations with military action. Argentina and Britain's long-standing territorial contest also meant they were fighting for more than oil in the *Shackleton* spat. Argentine naval officials, who planned the interception, also aimed to reassert their country's island claim and push the British to negotiate on the sovereignty issue.[36] In addition, the confrontation bolstered domestic and regional political support for the faltering Perón regime, as the Argentine population enthusiastically endorsed any initiative that challenged British authority in the islands and many Latin American governments supported Argentina's Malvinas claim.[37]

Like all oil spats, the *Shackleton* incident also remained quite mild. The *Storni* did not fire directly on the *Shackleton* or forcefully detain it, although its superior speed and firepower would have allowed it to do so. Consequently, the incident resulted in no material damage or injuries to sailors or civilians. Although the Argentine and British governments both protested the confrontation diplomatically, they did not escalate it militarily.[38] The Argentine government refused the navy's request to engage in "more drastic action," and the day after the confrontation, Rowlands insisted that the British government "shall do everything possible to cool the situation."[39] The states refrained from further militarized confrontations until 1982 and participated in no other oil-driven conflicts.

This restraint is indicative of countries' limited willingness to fight for oil. Although the Argentine government was prepared to intercept a British research ship to prevent it from exploring for oil in contested territories, it was not willing to invade the islands in order to grab their prospective petroleum resources. Doing so would have been easy, logistically; at the time, the Falklands were defended by thirty-seven lightly armed Royal Marines.[40] In addition, since the islands did not yet produce oil, invasion obstacles were nonexistent; there was no petroleum infrastructure to damage. Occupation obstacles would also be low, as a result of the islands' lack of oil facilities and small population. However, as the Falklands War and its aftermath later demonstrated, international and investment obstacles to aggression were high. Consequently, seizing the islands, for oil, would not pay.

Perpetrating an oil spat, in contrast, was worth the effort. Attempting to intercept a British research ship entailed no invasion or occupation obstacles, as the Argentines were not attempting to seize foreign territory. Investment obstacles were also irrelevant, as Buenos Aires was actively trying to discourage British resource development. Finally, international obstacles were minimal, as the incident was mild and occurred in contested territory. An oil spat was therefore a reasonable activity, especially since it would also advance other Argentine goals.

Following their oil sparring, Argentina and the United Kingdom did not intensify their competition over the Falkland Islands' prospective petroleum resources. Instead, they tried to use them as a basis for cooperation: a common response to oil spats. British officials had initiated this strategy before the *Shackleton* incident. In July 1975, Secretary Callaghan submitted a memorandum to the Overseas Policy and Defence Committee outlining a new approach to the contest, the objective of which was "to seek an interim settlement by using cooperation over oil on the Falkland Islands continental shelf and over the fishery resources of the South West Atlantic as constructive inputs for a new Anglo-Argentine dialogue."[41] Callaghan adopted this strategy for multiple reasons. First, the secretary recognized that it was the only way to develop the area's petroleum resources. As he put it, "With the difficulties over the continental shelf around the Islands it would have been difficult, if not impossible, for us to have done oil exploration or indeed oil production there without their [Argentine] cooperation."[42] Buenos Aires would resist any attempts at unilateral resource development and oil companies would be reluctant to participate in oil projects, because of the risk of Argentine intervention and the lack of British naval protection.[43]

Second, oil cooperation appeared to be the most viable means of managing the contest. The British government was not prepared to cede sovereignty over the Falklands to Argentina without the islanders' consent. However, the Foreign Office hoped that oil cooperation might act as a confidence-building measure. If the Falklanders collaborated with Argentina in resource exploration, they might develop greater trust and stronger ties with the mainland, which would increase their willingness to accept a change in political authority. Alternatively, oil cooperation might substitute for a sovereignty transfer. Prime Minister Harold Wilson hoped that Argentina would be willing to forgo the "legalities" of sovereignty if it could obtain the economic benefits, such as oil revenue.[44] At the very least, the government viewed the strategy as a means of kicking the can down the road. As long as Britain sustained negotiations, the Argentines would refrain from more forceful assertions of territorial sovereignty.

After the *Shackleton* incident, British authorities doubled down on oil cooperation. In 1977, Rowlands adopted a "mixed approach" to the dispute. His goal was to divide "sovereignty over people, and sovereignty over resources."[45] The Foreign Office would not relinquish authority over the islands without the Falklanders' consent. However, officials were willing to cede up to 90 percent of revenue from the Falklands' natural resources if that would persuade Buenos Aires that the British were not sustaining their grip on the islands because of oil and facilitate an agreement on political authority.[46] The Foreign Office hoped that, by establishing agreements on issues like oil cooperation, "at the outer edges of the problem," Britain could "provide a sufficient inducement to wean the Argentines away from their more extreme sovereignty claim."[47]

For their part, Argentine officials were open to the idea of joint resource development, as long as sovereignty remained on the negotiating table.[48] In June 1975, Foreign Minister Vignes had stated that he was "prepared to start talks on economic cooperation, with Argentina expecting a 50 per cent share."[49] The next year, following the *Shackleton* spat, the economic minister, José Alfredo Martínez de Hoz, endorsed the idea of oil cooperation.[50] In December 1977, Argentina and the United Kingdom agreed to continue their Falklands negotiations in two working groups: one on sovereignty and one on economic cooperation, including resource ventures.[51] Although the subsequent negotiations were rocky, some Argentine authorities remained optimistic about petroleum cooperation as late as summer 1980. In June of that year, when asked about the ongoing negotiations, Martínez de Hoz stated, "Some progress has been made and there is a little light on the horizon . . . and I think the economic side could help. We have two common interests, which could be oil and fishing. So long as some sort of discussions on sovereignty can go on at the same time we might be able to reach some kind of agreement on joint oil exploration or fishing which would be the beginning of a get-together on this issue."[52]

The *Shackleton* spat, while causing an uproar at the time, did not significantly worsen Anglo–Argentine relations; it was one, fairly minor incident in a larger, longer dispute. Authorities in both states endeavored to prevent the conflict from escalating, and the oil spat encouraged further efforts to collaboratively develop petroleum resources. Thus, while the confrontation was oil-driven, it was also quite benign. As a result, relative to the many other issues pushing the states toward war in 1982, the 1976 confrontation played a very minor causal role.

The Falklands War (1982)

Despite Martínez de Hoz's positive assessment, by late 1980, prospects for an Anglo–Argentine agreement, on either oil cooperation or sovereignty, were fading. As the British Parliament repeatedly rejected political proposals, Argentine officials became increasingly convinced that their opponents were negotiating in bad faith: that they had no intention of ever voluntarily relinquishing island sovereignty. In January 1982, the Argentine military began to draft new plans for a Falklands invasion as a fallback if talks failed.[53] This planning accelerated in March 1982, following confrontations on South Georgia Island, a British dependent territory, 860 miles southeast of the Falklands. That month, an Argentine scrap merchant, Constantino Davidoff, traveled to South Georgia to salvage metal from abandoned whaling stations. He informed British authorities of the plan before his departure and received no opposition. However, when his party arrived, members of the British Antarctic Survey, stationed on the island, warned that the workers might have been accompanied by military personnel. Suspecting that the operation was a cover for Argentine occupation, the British government threatened to forcibly evict Davidoff and his men.[54]

The intensity of this response shocked Argentine authorities. They were also agitated by reports of a British nuclear submarine sailing south, toward the Falklands. Once the vessel arrived in the region, Argentine forces would no longer be able to conduct surface operations around the islands or land at the capital, Stanley. Fearing that their window of opportunity for retaking the Falklands was closing, the Galtieri government decided to act.[55] It deployed a naval task force, which invaded the islands on April 2. The Thatcher government retaliated by deploying its own task force, which forcibly retook South Georgia and the Falkland Islands over the next two months in battles that resulted in over one thousand military fatalities. To Argentine leaders, the magnitude of Britain's response came as an unwelcome surprise. They had expected that their aggression would provoke an intensification of diplomatic negotiations, not an interstate war.[56]

Some academic researchers and popular commentators have labeled the 1982 Falklands conflict a classic oil war.[57] However, this characterization is inaccurate. Argentina did not launch the war in order to seize petroleum resources. By the early 1980s, enthusiasm for the Falklands' oil prospects had waned. The Shackleton survey's report, issued in July 1976, was not particularly optimistic about the region's oil potential.[58] In addition, by summer 1978, both governments had begun to receive the results from further seismic surveys of the continental shelf between the islands and the Argentine mainland.[59] Although the reports were not made public, oil companies that purchased them commented on their content. One Argentine oil expert stated that the results were "interesting," while others observed that they "[do] not rule out the presence of hydrocarbons." However, they concluded that, overall, "prospects are less enticing than once believed."[60] This disappointed both governments, which had hoped for positive findings, as "by and large, it was supposed, the more likely it was that oil was present, the greater the possibility for some sort of deal."[61]

Argentine officials, in particular, were unimpressed by the Falklands' oil prospects. A British Joint Intelligence Committee report observed that Argentines regarded the islands' petroleum potential as "small and very long term."[62] Similarly, Guillermo Zubaran, Argentina's minister of energy, stated in 1977 that he found the reports issued by the US Geological Survey and Shackleton expedition "excessively optimistic."[63] In 1982, Argentine authorities had even less interest in the Falklands' resources because, by then, they had plenty of oil at home. The state's domestic petroleum output had grown rapidly in the 1970s, causing Argentina to become a net oil exporter in 1980.[64] The country could also continue to expand its petroleum production without the Falklands, because the richest areas of its continental shelf were believed to lie closer to the mainland.[65] Finally, if the Argentines wanted the Falklands' oil, it was available to them. As noted earlier, British authorities were willing to grant Argentina a large share of the Falklands' resource revenue if that would facilitate a political agreement that respected the islanders' wish to remain attached to the United Kingdom. Argentine officials did not pursue this course of action because, while they possessed mild interest in the

Falkland's prospective petroleum resources, their desire to regain island sovereignty was paramount.[66] Contrary to Prime Minister Wilson's hopes, no amount of progress on the former would substitute for the latter.

Argentine officials' determination to retake the Falklands, coupled with their increasing conviction, after a quarter century of failed negotiations, that a transfer could only be accomplished through military force, was the primary cause of their invasion. The Galtieri government also felt a mounting sense of time pressure. The regime was determined to recover the Falklands before January 3, 1983: the 150th anniversary of Britain's occupation. The military junta feared that, if they failed to accomplish the sovereignty transfer by then, internal opposition would be intense.[67] Hence, the Falklands War was partly precipitated by domestic political concerns, although not the diversionary mechanism that is often associated with the conflict.[68] Like many long-standing territorial contests, the war was also intimately intertwined with national pride. The Argentine people and their government wanted to correct what they perceived as an unacceptable "historical injustice."[69]

The United Kingdom's decision to retaliate for Argentina's invasion was also unconnected to oil ambitions.[70] Britain's petroleum production, like Argentina's, had boomed between 1976 and 1982. During this time, the state's North Sea output grew from nothing to over two million barrels per day, transforming it into a net oil exporter. The British had little interest in obtaining more oil, especially resources that were uncertain in volume, politically provocative, and expensive to exploit because of their challenging physical location.[71] Officials did hope that a small percentage of revenue from the Falklands' eventual oil exploitation could be directed toward the islands' economic development to lessen the British government's financial burden.[72] However, as noted, they were willing to cede most resource revenue to Argentina if that would facilitate a political settlement. The Thatcher government fought out of a sense of national pride, to demonstrate that Britain remained a meaningful military power, to resist international aggression, and to avoid antagonizing the British public, which demanded a forceful response to the invasion.[73] It did not fight for oil resources.

The Falkland/Malvinas Islands dispute demonstrates that states are willing to initiate small confrontations to advance or defend their petroleum claims, particularly when they are engaged in long-standing territorial disputes and can anticipate additional gains from petroleum sparring, such as rallying domestic support. Under those circumstances, minor, low-cost military actions, like the *Shackleton* spat, are worth the effort. Classic oil wars, in contrast, do not pay. Recognizing this, the Argentine government did not launch the Falklands War to seize the islands' prospective petroleum resources. Instead, like the Chaco and Iran–Iraq conflicts, the Falklands War was a red herring.

7

OIL CAMPAIGNS

World War II

Like oil spats, oil campaigns are prosecuted primarily to obtain petroleum resources. Unlike oil spats, however, these are severe episodes that result in hundreds or thousands of fatalities. Nonetheless, suggesting that the larger conflicts in which these campaigns occur are classic oil wars, as many authors do, is misrepresentative.[1] Aggressors in oil campaigns do not initiate wars to seize petroleum resources. Instead, they launch their oil-oriented attacks in the midst of ongoing conflicts that were started for other reasons. These existing wars heighten aggressors' petroleum needs by increasing their oil consumption and limiting their access to foreign resources. Eventually, obtaining more oil becomes a national security imperative. Still, resource-hungry states hesitate to launch oil campaigns. Recognizing the obstacles to grabbing petroleum, they exhaust all other means of satisfying their oil needs before turning to international aggression. Many also initiate their oil campaigns reluctantly, aware that seizing foreign petroleum likely will not pay, even in wartime. Their hesitation suggests that, in the absence of existing conflicts, the aggressors would have refrained from fighting for oil. Moreover, even in oil campaigns, petroleum is not the sole issue that states are fighting for.

This chapter examines the two most prominent historical oil campaigns: Japan's invasion of the Dutch East Indies and northern Borneo (1941–1942) and Germany's attacks against the Soviet Union (USSR) (1941–1942). These cases are regularly identified as classic oil wars; Japan and Germany purportedly attacked their targets in order to grab petroleum resources. This interpretation is not entirely inaccurate; German and Japanese leaders did initiate these specific campaigns largely to obtain foreign oil. However, it is also incomplete, as it fails to acknowledge that these attacks occurred during ongoing wars: specifically, the Second Sino–Japanese War (1937–1945) and World War II in Europe (1939–1945). Neither of these existing conflicts was a classic oil war. Instead, Japan and Germany aspired to become the hegemonic powers in their respective geographic regions.

As the wars continued, however, both belligerents' energy insecurity mounted. Neither Japan nor Germany possessed large domestic petroleum endowments. Meanwhile, prewar militarization and early acts of territorial aggression increased national oil consumption while undermining the states' access to foreign crude and petroleum products. Recognizing the obstacles to seizing foreign resources, both countries initially attempted to avoid oil campaigns. Instead, they responded to their energy crises with nonconquest strategies: developing their limited domestic oil resources and petroleum substitutes, stockpiling, trading, bargaining, and, in Germany's case, even engineering a nonviolent occupation of a neighboring state to satisfy national petroleum needs. However, each of these alternative strategies eventually proved to be inadequate. Authorities were therefore faced with a choice: grab foreign oil or lose their ongoing war. At that point, both states turned to oil campaigns. Germany attacked the USSR in Operation Barbarossa (1941) and Case Blue (1942), advancing toward oil fields in the Caucasus. Japan invaded the petroleum-rich Dutch East Indies, British protectorates in northern Borneo, and Brunei (1941–1942). It also preceded the invasion with an attack on the United States' naval base at Pearl Harbor, partly to prevent US forces from retaliating for the southern attacks.

Yet even under these desperate circumstances, the two countries' oil campaigns were not driven exclusively by petroleum ambitions. Hitler's hostility toward the USSR was long-standing, and in Operation Barbarossa, he aimed to defeat Germany's rival for European hegemony, eliminate the perceived threat of Bolshevism, obtain more lebensraum (living space) for the German people, drive the United Kingdom out of the war, and prevent the United States from entering it, as well as seize oil resources. Case Blue was more petroleum focused but also included an attack on Stalingrad. Notably, Germany initiated the latter oil campaign, in which petroleum was a dominant goal, less optimistically. Meanwhile, Japan attacked Pearl Harbor and the Philippines partly to safeguard its oil access, but also because, over the previous four decades, its leaders had come to view the United States as an irremediable adversary that was determined to prevent Japan's regional rise. The Japanese launched their campaign in order to escape from this perceived "encirclement," as well as to acquire more petroleum. These added incentives, coupled with the aggressors' delays in launching their oil campaigns and these attacks' timing, in the midst of ongoing wars, support the book's larger argument: that states avoid fighting for petroleum resources.

Japan's Invasion of the Dutch East Indies (1941–1942)

Japan possesses such limited domestic petroleum endowments that, in the 1930s, the state extracted less oil in a year than the United States, the world's leading petroleum producer, did in a day.[2] Consequently, Japan relied on foreign imports for 93 percent of its domestic oil consumption. Over 80 percent of its imports came from the United States, while another 14 percent were supplied by the Dutch East Indies.[3] Japanese authorities, like all consumer state leaders, would have preferred to be less reliant on foreign oil, particularly after World War I revealed the resource's importance for modern warfare.[4] However, Japan did not initiate an oil campaign until the winter of 1941–1942, when it attacked the Dutch East Indies and northern Borneo.[5]

Japan's oil campaign occurred during the Second Sino–Japanese War, widely known as the China Incident. This conflict precipitated Japan's oil campaign by increasing national petroleum demand and constraining the state's ability to purchase foreign oil resources. The Second Sino–Japanese War, however, was not caused by petroleum ambitions. Instead, it arose from Japan's desire to become the dominant power in East Asia.[6] This aspiration had emerged in the late nineteenth century among Japanese elites who observed that their country was rapidly transforming from a traditional agricultural state into a modern industrialized power.[7] It was cemented in 1905, when Japan defeated Russia, its main regional rival, in the Russo–Japanese War (1904–1905). The victory sent shockwaves through the international community, as it was the first time that an Asian country had defeated an established, European great power. It also convinced the Japanese that their country was destined for regional leadership.

Over the next four decades, Japan extended its territorial authority in East Asia. The Treaty of Portsmouth (1905), which settled the Russo–Japanese War, gave the state control over Port Arthur, on China's Liaodong Peninsula, and the southern half of Russia's Sakhalin Island. Over the next decade, Russia and Japan partitioned Manchuria, a semiautonomous territory in northeastern China, into separate spheres of influence. In 1910, Japan annexed Korea. Five years later, Tokyo exploited the instability created by World War I to extend its "special rights" in Manchuria, including expanding its participation in the area's resource development.[8] Japan invaded the region after the Mukden Incident, in September 1931, when Kwantung Army officers deliberately destroyed part of the South Manchuria Railway and blamed Chinese troops for the attack. The central government had not ordered the Mukden operation but subsequently embraced it after observing its domestic popularity. Tokyo proclaimed Manchuria's independence from China, renamed the province Manchukuo, and installed a puppet government. In early 1933, Japan invaded the neighboring province of Jehol and annexed it to Manchukuo.[9]

The Second Sino–Japanese War began in July 1937, when Japanese and Chinese forces exchanged fire along the Marco Polo Bridge, near Beijing. The incident was preceded by multiple years of Sino–Japanese friction over authority in north China.[10] Nevertheless, the confrontation was mild and it initially appeared that it could be resolved peacefully. Most Japanese leaders did not seek a war, particularly one that extended beyond China's northern provinces; neither the civilian government nor the army general staff had any interest in acquiring formal control over central or southern China.[11] However, some army factions were eager to pursue more forceful action, and their early victories encouraged further aggression. In addition, the Japanese navy, convinced that the conflict would spread to Shanghai, persuaded cabinet officials to deploy army divisions to the city and to Tsingtao (Qingdao). The Chinese Nationalists, led by Chiang Kai-Shek, also escalated the confrontation. Consequently, by mid-August, China and Japan were engaged in an all-out war.[12] The conflict would persist until Allied forces defeated Japan in 1945.

The outbreak of the Second Sino–Japanese War was therefore "more a product of drift than of policy."[13] However, this expansionism was consistent with Japan's broader ambition of becoming the dominant power in East Asia. In the 1930s, Japanese leaders issued a variety of policy proposals articulating that goal, including the Amau Declaration (1934), which effectively proclaimed a Monroe Doctrine for East Asia, with Japan in the hegemonic role; the New Order for East Asia (1938), a collective security and economic system involving Japan, Manchukuo, and China; and the Greater East Asia Co-prosperity Sphere (1940), which extended the New Order concept to French Indochina and the Dutch East Indies. Politically, these initiatives aimed to defend Japan from communist threats and strengthen its regional standing relative to the USSR, United States, and United Kingdom. Economically, they sought to secure Japan's access to vital raw materials and create preferential markets for Japanese goods. Leaders believed that creating a regional economic bloc was the only way for a "have-not" country like Japan, which lacked large internal markets and raw material endowments, to compete against "have" countries like the United States and USSR.[14]

None of Japan's early acts of expansion, however, aimed to seize foreign oil. Korea did not produce petroleum and was valued primarily for its food supplies.

Manchuria was also prized predominantly for its food production and coal and iron resources, not oil.[15] The province did possess some oil shale reserves, at Fushun, which were being exploited by 1931. However, converting oil shale into usable petroleum products was a highly inefficient process and, even with large navy subsidies, Fushun's output never fulfilled more than 5 percent of Japan's oil needs.[16] Capturing Jehol and other parts of northern China also failed to augment Japan's oil endowments; neither had significant petroleum prospects. Nor did the rest of the country. Although oil had been discovered in Shensi Province in the 1910s, these finds were so limited that foreign oil companies withdrew from the region in 1915.[17]

Rather than significantly enhancing Japan's oil access, the state's early expansionism undermined its energy security. The intrusions into China, in particular, dramatically increased Japan's petroleum needs.[18] In addition, the state's rapid industrialization created a surge in domestic petroleum use; from 1931 to 1937 alone, national oil consumption doubled.[19] To satisfy this demand, Japan had to increase its imports of crude oil and petroleum products. Paying for these foreign resources drained the state's gold and foreign exchange reserves, thereby endangering its long-term ability to purchase raw materials.[20] If this pattern persisted, Japan would eventually be unable to acquire sufficient oil resources to power its military and industrialized economy, causing it to "drop back to a poorer standard of life, and a lesser rating among the powers."[21]

The Second Sino–Japanese War also undermined Japan's energy security by provoking international retaliation from third-party states and international organizations. In January 1932, US secretary of state Henry Stimson responded to Japan's invasion of Manchuria with a formal note asserting that the United States did not recognize Japanese authority in the region and lambasting Tokyo for violating the Nine-Power Treaty (1922), which guaranteed all of the great powers equal access to trade in China. In addition, Stimson warned that, if Japan continued to disregard its international obligations, other signatories might also ignore their treaty commitments. In March, at the United States' urging, the League of Nations endorsed Stimson's "non-recognition" doctrine.[22]

International responses to the subsequent Second Sino–Japanese War were even harsher. In October 1937, US president Franklin D. Roosevelt delivered his "Quarantine Speech," which identified Japan as a "bandit nation" for its aggression in China. In July 1938, the United States responded to Japan's bombing of Canton by implementing a "moral embargo," which asked US aircraft manufacturers to voluntarily restrict airplane and engine sales to Japan in order to limit the country's warfighting capabilities. In September 1939, the moral embargo was extended to the materials and technological information required to manufacture high-octane aviation gasoline.

In July 1939, the Roosevelt administration also announced its withdrawal from the United States' 1911 commercial treaty with Japan, effective six months later. By terminating the agreement, US authorities removed all legal impediments to formal trade embargos against Japan. Robert Craigie, the United Kingdom's ambassador to Tokyo, stated that the renunciation was met with "considerable shock" by Japanese leaders, who perceived it as a major threat to national security, particularly because of its potential impact on US oil exports.[23] In July 1940, the United States began to exploit its new liberty by implementing the Export Control Act, which enabled the government to limit sales of war materials to Japan. By August 1940, the Roosevelt administration had restricted exports of aviation fuel and lubricants. In September, it responded to Japan's occupation of northern Indochina and membership in the Tripartite Pact with Germany and Italy by restricting iron and steel scrap exports. From February to March 1941, the United States limited exports of oil drilling and refining equipment, along with metal oil drums, in order to constrain Japan's ability to produce and transport petroleum products.[24]

These trade restrictions concerned Japanese leaders partly because they threatened the state's oil access. However, authorities were also worried because they viewed the export controls as the latest expression of a long-standing American foreign policy goal: to prevent Japan's rise as a great power.[25] This aspiration, Japanese officials believed, had manifested itself in multiple ways since the late nineteenth century. First, it appeared in the states' competition for political and economic authority in the Pacific and East Asia. In 1898, after years of Japan and the United States jockeying for influence in Hawaii, the McKinley administration formally annexed the territory. The same year, the United States' victory in the Spanish–American War gave it control over the Philippines and Guam, bringing US forces closer to Japan and further impinging on Tokyo's desired sphere of influence.[26] In addition, the United States pushed for an open trading system in China rather than granting Japan a privileged position in the neighboring country.

Second, successive American administrations restricted Japanese immigration to the United States. In the early twentieth century, many Japanese viewed emigration as key to their country's development. It relieved rising population pressures at home and provided economic opportunities for Japanese citizens living abroad, who would then transfer wealth back to the home islands. However, individuals who moved to the United States faced severe discrimination, including school segregation and restrictions on land ownership. By 1908, the United States was impeding the immigration of Japanese laborers. Then, the Japanese exclusion provision of the 1924 Immigration Act terminated immigration entirely.[27] The racist overtones of these policies were not lost on the Japanese, who also blamed the United States for blocking a racial equality clause in the covenant of the League of Nations (1919), which Japan had lobbied for at the Paris Peace Conference.[28] The hostility engendered by the immigration issue was so intense that it led to a brief war scare in 1906 and, in the aftermath of World War II, Emperor Hirohito cited it as one of the conflict's fundamental causes.[29]

Third, Japanese officials' suspicions were heightened by US hostility to their regional military dominance. At the London Naval Conference (1930), Secretary Stimson insisted on maintaining the ten-to-six ship ratio between the United States and Japan that the countries had agreed to at the Washington Naval Conference (1921–1922). When Japanese negotiators attempted to revise the ratio upward to ten-to-seven, which they felt was the minimum balance necessary for guaranteeing their naval superiority in the Pacific, Stimson threatened an Anglo–American agreement excluding them. Japan's representatives could see no reason for this intransigence "unless the United States actually harbored the idea of ultimately engaging in a war with the Japanese empire."[30] US officials were equally inflexible during talks on revising the ratio in late 1934, heightening Japanese concerns.[31]

Lastly, the Japanese resented US trade barriers. In response to the Great Depression, the United States, like many other countries, including Japan, imposed increasingly large tariffs on foreign imports. The Smoot–Hawley Act (1930) and subsequent protectionist measures reduced American purchases of inexpensive Japanese manufactured goods, causing the share of Japan's exports flowing to the United States to drop by almost 40 percent in the early 1930s.[32] This collapse occurred at a particularly inopportune moment for Japan, as the price of raw silk, one of its leading exports, had just collapsed, prompting the country to seek new markets for other export products to compensate for the revenue loss.[33] US trade restrictions therefore contributed to Japan's persistent balance-of-payments problem. They also undermined leaders' faith in the global trading system and strengthened their commitment to developing a regional economic bloc.[34]

In 1931, Japanese intellectual Tsurumi Yusuke summarized the United States' purportedly anti-Japanese actions by observing that the country had erected "three unsurmountable walls: the tariff wall, the emigration wall and the peace wall. The first wall excludes Japan's manufactured goods from other countries. The second cuts off the migration of her people. And the third prohibits the readjustment of the unequal distribution of territories among nations with different density of population."[35] Tsurumi's sense of aggravation was widely shared. Arriving in Tokyo in June 1932, the US ambassador to Japan, Joseph C. Grew, reported "a tornado" of anti-American hostility.[36] Although tensions declined by the next summer, the United States' territorial, immigration, and trade policies created a reservoir of animosity that the Japanese could draw on whenever they were confronted by renewed American obstructionism. As Grew stated at the time, "Latent hostility towards the United States is always present and any provocation would fan it into flame."[37] US naval exercises near Midway in the spring of 1935 reawakened Japanese resentments and apprehension, as did the stationing of much of the US fleet at Pearl Harbor.[38]

Japanese suspicions mounted further during the Second Sino–Japanese War. Americans had favored the Chinese before the conflict, and their sympathies deepened after Japan's unprovoked sinking of the USS *Panay* on the Yangtze River and the brutal Rape of Nanking, from December 1937 to January 1938.[39] During the summer of 1938, the Roosevelt administration began to financially support Chiang Kai-Shek's Nationalist forces.[40] The United States offered additional loans to China in 1940 and, that October, Britain reopened the Burma Road, a Nationalist supply route.[41] In May 1941, the United States extended lend–lease assistance to China, providing weapons and other war materiel through Burma and India. By autumn, the US government had begun to train, supply, and even staff part of China's air force.[42] The Japanese viewed this assistance as both a threat to their national security and further evidence of the United States' determination "to deny Japan her rightful place in the world."[43]

Japan's expansion into Southeast Asia was precipitated partly by this foreign assistance. Japanese officials attributed Chiang's unwillingness to surrender to Western support; they believed that, as long as the Nationalists continued to receive outside aid, they would sustain their resistance, preventing Japan from bringing the Sino–Japanese conflict to a favorable close. Officials also concluded that, to interrupt foreign assistance, Japan must target the territories that contained international supply routes. The Imperial Navy invaded Hainan in February 1939, partly to obtain bases for bombing international supply lines. Japanese forces then occupied northern Indochina in September 1940 and advanced into southern Indochina in July 1941. Supply interdiction was one of their goals. However, the Japanese also aimed to acquire raw materials for the ongoing Sino–Japanese war and create a bulwark against "encirclement" by the "ABCD" powers (American, British, Chinese, and Dutch).[44] In addition, the situation in Europe offered a "golden opportunity" for Japanese expansion; as Ambassador Grew wrote, "The German victories have gone to their heads like strong wine."[45]

Yet Japan did not seize this opportunity to grab foreign oil resources.[46] Instead, the state attempted to increase its energy security nonviolently. Since the mid-1930s, many Japanese had advocated closer ties with the Dutch East Indies and northern Borneo, partly to enhance national resource access. The navy, in particular, embraced southern expansion as a means of "strengthening our national defense and solving the population problem and economic development."[47] In 1939, the Navy National Policy Research Committee outlined a program for the

MAP 7.1. Southeast Asia (1939)

south, aimed at securing raw materials, promoting Japanese exports and business endeavors, and developing the region's agricultural and fisheries industries with Japanese participation. However, the plan did not advocate a conquest strategy to achieve these goals. Instead, it sought to expand southern engagement through "peaceful means," including strengthening commercial, cultural, and diplomatic ties and, in the Dutch East Indies, establishing a cultural agreement or nonaggression treaty.[48] Although popular enthusiasm for a forceful occupation mounted after Germany's rapid victories in western Europe, military authorities maintained this more restrained stance. The army and navy's "General Principles to Cope with the Changing World Situation" (July 1940) stated that the Dutch East Indies' resources should be secured only through "diplomatic measures."[49] The services' subsequent "Outline of Policy towards the South" (April 1941) asserted that Japan should not resort to military force unless "no other means are available," and then only "for the sake of its self-existence and self-defense."[50]

This circumspect attitude did not arise from military constraints; if Japan invaded the Dutch East Indies and British protectorates, it would easily conquer them. Colonial forces were insufficient to resist Japanese aggression, especially after Germany's attacks on the Netherlands and United Kingdom.[51] However, the Japanese recognized that a southern oil grab would face severe invasion obstacles. In the event of a Japanese attack, the Dutch and British companies that managed local petroleum installations would attempt to destroy their facilities.[52] If this self-sabotage incapacitated local wells and refineries, Japan's petroleum payoffs would be temporarily limited; although it would control oil resources, it would not be able to exploit them.

These invasion obstacles could be rectified over time. However, the greater impediment to seizing foreign oil was the threat of international retaliation. In May 1940, Japanese naval war games revealed that, if the state attacked the Dutch East Indies, the United States would respond militarily. Recognizing these severe international obstacles, most naval officers were opposed to seizing the region's oil. When these findings were communicated to the army in February 1941, they also dampened that service's enthusiasm for a forceful southern advance.[53] High-ranking military officials recognized that, although Japan might achieve short-term victories in the South Seas, it was unlikely to triumph in a protracted war with the United States.[54] As Chief of the Naval General Staff Nagano Osami told the emperor in July 1941, "A great victory like the Battle of the Japan Sea is out of the question, and I am even uncertain whether we can win or not."[55] Admiral Yamamoto Isoroku, who led Japan's attack on Pearl Harbor, was even more pessimistic, asserting that a war with the United States would be the "height of folly" and stating, "We shall not be able to stand up to them."[56] If Japan was defeated, it would be compelled to withdraw from the Dutch East Indies and northern Borneo, relinquishing all petroleum payoffs.

Officials also recognized that petroleum payoffs would be compromised well before Japanese forces were pushed out of the occupied territories. Although the Imperial Navy could consume some of the British and Dutch East Indies' petroleum locally, once regional refineries were repaired, much of the area's oil would need to be transported to the Japanese home islands. This transit route was 3,500 miles long, and Japanese authorities assumed that the United States would attack oil tankers, using submarines or bombers deployed from bases in the Philippines.[57] Even if Japan seized these military installations, planes launched from

US aircraft carriers could intercept oil shipments once they were within a 200-mile range.[58] Moreover, in 1941, Japan's tanker fleet was already insufficient to transport needed petroleum supplies to the home islands and wartime losses would exacerbate the shortage.[59] These international obstacles, coupled with the invasion obstacles that Japan would face immediately after seizing Borneo and the Dutch East Indies, rendered an oil campaign an unappealing enterprise.

To avoid fighting for oil, Japanese leaders initially adopted a variety of alternative strategies to satisfy national petroleum needs. First, they pursued overseas oil concessions, which would give them direct control over resource exploration, production, and sales, thereby enabling the government to direct foreign petroleum supplies to the home islands. In 1925, Japan successfully obtained a concession from Russia on northern Sakhalin Island, which supplied the country with over seven hundred thousand barrels of oil annually. However, subsequent efforts to acquire concessions in Mexico, Ecuador, Romania, Burma, and northern India failed.[60]

Second, Japan stockpiled oil resources. In 1934, Tokyo established the Petroleum Industry Law, which required foreign oil companies operating in Japan to maintain a stockpile equivalent to six months of their normal imports. The state also developed its own petroleum stocks by increasing foreign oil purchases. However, this practice exacerbated Japan's foreign exchange crisis and was still vulnerable to supply cutoffs from petroleum-producing states.[61] Consequently, Japan also employed a third strategy: attempting to develop oil substitutes. The state established a Fuel Bureau in May 1937 and tasked it with promoting synthetic fuels production. The program aimed to fulfill half of Japan's oil needs by transforming coal resources into gasoline and other fuels using liquefaction and gasification processes. However, these efforts were a dismal failure. At its peak in 1944, the program met only 11 percent of its output target. In other years, its performance was even worse.[62]

Fourth, Japan attempted to secure additional oil supplies from the Dutch East Indies. In February 1940, the Japanese Foreign Office requested talks with the Netherlands on oil purchases. The Dutch did not initially reply. However, they were more responsive when Japanese officials repeated their request in May, after Germany invaded the Netherlands. Later that month, authorities acceded to Japan's demand to double its annual oil purchases from ten thousand to twenty thousand barrels per day.[63] Still viewing this amount as insufficient, Japanese officials secured Dutch permission to send a trade delegation to Batavia (Jakarta). When the Kobayashi mission arrived in September 1940, it requested oil concessions in the Dutch East Indies and over twenty-two million barrels of petroleum per year: a threefold increase over the May agreement. The Dutch demurred on concessions but, in October, Royal Dutch Shell and the Standard–Vacuum Oil Company offered Japan six- to twelve-month contracts for thirteen million barrels of crude oil and distillates.[64]

The Japanese took the offer but, still unsatisfied, deployed another trade delegation, led by Yoshizawa Kenkichi, in late December. This initiative was more comprehensive; the Japanese essentially aimed to incorporate the Dutch East Indies into their regional economic bloc. However, Batavia's resistance had stiffened; the colonial government imposed new controls on raw material exports, offered Japan only small quantities of resources, and dragged its heels in negotiations. In June 1941, Yoshizawa was recalled. All that Japan had achieved during the mission, in terms of oil, was the renewal of the existing six-month contracts.[65] Frustrated Japanese leaders blamed the United States for these diplomatic failures. They believed, accurately, that the Roosevelt administration had pressured the local government and oil company representatives to resist Japanese demands.[66]

Japan's final efforts to peacefully obtain additional oil resources targeted the United States. In early 1941, the two countries initiated the Hull–Nomura talks, named for their primary participants: US secretary of state Cordell Hull and Japan's ambassador to the United States, Nomura Kichisaburo. Their early discussions did not tackle the oil issue specifically; instead, they aimed to contain escalating Japanese–American tensions. During the first half of the year, the talks made little progress, partly because of Japanese aggression and partly because of American officials' lack of flexibility. Hull had insisted from the outset that any bilateral accord be premised on Japan's acceptance of the "Four Principles": respect for the territorial integrity and sovereignty of all nations, noninterference in the internal affairs of other countries, equality of commercial opportunity, and nondisturbance of the status quo in the Pacific, except by peaceful means. The Americans also demanded Japan's public renunciation of the Tripartite Pact and its full withdrawal from China.[67]

These conditions were anathema to Japanese leaders, who viewed the maintenance of their sphere of influence in East Asia as vital for national defense and raw materials acquisition. Withdrawing from China would also force Japan to conclude the Second Sino–Japanese War with no gains: an outcome that was believed to be unacceptable to the Japanese public after four years of conflict.[68] At a meeting of Japan's leaders, Foreign Minister Togo Shigenori claimed, "For the United States to insist that Japan disregard the sacrifices she is making in China is tantamount to telling us to commit suicide."[69] Were Japan to retreat from China, he asserted, "the international position of our Empire would be reduced to a status lower than it was prior to the Manchurian Incident, and our very survival would inevitably be threatened."[70] Moreover, although Japanese officials were willing to quietly ignore their state's commitments under the Tripartite Pact, a public breach with Germany was deemed unacceptable. As President of the Privy Council Hara Yoshimichi summarized, "It is impossible, from the standpoint of our domestic political situation and of our self-preservation, to accept all of the American demands."[71]

Officials were nonetheless more accommodating after July 25, 1941, when the Roosevelt administration responded to Japan's invasion of southern Indochina by freezing all Japanese assets in the United States. This action was not initially intended to halt US oil exports; the order contained provisions for licensing petroleum transactions. However, the procedures were never implemented, so the freeze became a de facto oil embargo. Since the Dutch and British followed suit, by September, no foreign oil was reaching Japan. In the wake of these restrictions, the state's petroleum stockpiles began to decline at the rate of eighty-four thousand barrels per day.[72] Although Japan had built up two years of stocks, this rapid drop raised the specter of future petroleum shortages. Without access to additional oil supplies, Japan would not be able to power its military or domestic industries. It would be forced to capitulate in China and would become vulnerable to American and Soviet attacks. Domestic shortages of energy resources would also lead to popular discontent. Should these conditions continue, Togo predicted to Ambassador Grew, Japan "would inevitably collapse," through either outside aggression or internal opposition.[73]

In the wake of the oil embargo, Japanese leaders pushed for a resumption of the Hull–Nomura talks. They also attempted to arrange a personal meeting in Hawaii between President Roosevelt and Japan's prime minister, Prince Konoe Fumimaro. Such a move was unprecedented for a sitting Japanese leader, and moderate Japanese officials hoped that, if the president and prime minister met abroad, Konoe would be able to compromise further, negotiating a settlement that could be presented to the more aggressive members of the Japanese leadership as a fait accompli.[74] However, the State Department stalled on meeting preparations and, eventually, the proposal was scrapped. Japanese officials began to suspect that they "had fallen into a trap" and that "the United States never had any intention of coming to any agreement."[75] Nonetheless, the Tojo Hideki government, which succeeded the Konoe government in mid-October, persisted in negotiations, even with the knowledge that delaying an attack would further compromise Japan's chances of victory.[76]

On November 7, 1941, increasingly desperate Japanese officials submitted another proposal, representing the most they believed they could concede in response to American demands without threatening their state's existential security. Concessions included withdrawal from most of China within two years of a resolution of the Sino–Japanese war and withdrawal from Indochina as soon as the conflict concluded. In addition, the proposal accepted the principle of nondiscrimination in trade and included strong indications that Japan would not be bound by the Tripartite Pact.[77] However, rather than responding with a viable counteroffer, Hull replied on November 26 with the Ten Points Plan: an even more forceful and expansive articulation of American demands. The secretary also asserted that the United States would make no further compromises in its position. This ultimatum terminated negotiations; as historian Paul Schroeder observes, "Even the most moderate and conciliatory Japanese regarded the American terms as completely unacceptable and humiliating."[78] Hull himself recognized that his response would torpedo bilateral talks; in his diary, he recorded that he had decided "to kick the whole thing over."[79]

With the possibility of a diplomatic settlement scuttled, Japan reluctantly turned to international aggression. Military and civilian leaders recognized the substantial obstacles to an oil campaign. They knew that local oil companies would sabotage their facilities and that attacking the Dutch East Indies would provoke a forceful US response. Yet, after exhausting all other means of meeting national petroleum needs, they felt that they had no other choice. As Tojo stated, "Under the circumstances, our Empire has no alternative but to begin war against the United States, Great Britain, and the Netherlands in order to resolve the present crisis and assure survival."[80] Although the endeavor was unlikely to succeed, "rather than await extinction it were better to face death."[81] On December 16, 1941, Japan invaded Brunei and the Kingdom of Sarawak, the oil-producing territories in northern Borneo. The state invaded Dutch Borneo in January 1942 and Sumatra in February. In addition, Japanese forces attacked the American naval base at Pearl Harbor and the Philippines on December 7–8. Despite joint resistance from British, Dutch, Australian, and American forces, by early March, Japan controlled the entire southern region, including key petroleum facilities at Balikpapan, in southern Borneo, and Palembang, on Sumatra.

Gaining control over foreign oil was a central incentive for Japan's attacks on the Dutch East Indies and northern Borneo. Yet these oil aspirations did not arise on their own. Instead, they emerged from Japan's ongoing war with China, which increased the state's oil demand and constrained its ability to purchase foreign petroleum because of the opposition its expansionism provoked from third-party states. The Second Sino–Japanese War, however, was not launched for oil. And, in its absence, Tokyo would not have attempted the "gigantic gamble" of seizing foreign petroleum.[82] In addition, oil ambitions were not the only factor motivating Japan's oil campaign. Authorities also believed that their country faced a broader, existential threat from the United States. Consequently, they interpreted the oil embargo as merely the latest, most immediately menacing action in a long history of the United States employing "all available means" to reduce Japan to "a third-rate country."[83]

Based on the states' long, acrimonious history, Japanese officials concluded that American opposition would be unrelenting; as Konoe stated at a leaders' conference on October 28, 1941, "Even if we should make concessions to the United States by giving up part of our national policy for the sake of a temporary peace, the United States, its military position strengthened, is sure to demand more and more concessions on our part; and ultimately our empire will lie prostrate at the feet of the United States."[84] Japan launched its oil campaign not only to acquire petroleum but also to avert this outcome. Leaders were uncertain about the results, recognizing that they were unlikely to win a prolonged war against the United States or maintain long-term control over foreign petroleum resources. Nonetheless, they preferred an oil campaign to certain defeat.

Germany's Invasion of the USSR (1941–1942)

Germany's trajectory in the 1930s and early 1940s was very similar to Japan's. Like Japan, Germany lacks significant domestic petroleum endowments. However, having recognized the obstacles to seizing foreign petroleum, the state refrained from launching an oil campaign until June 1941, when it invaded the USSR in Operation Barbarossa. Before initiating this attack, the German government exhausted all alternative means of satisfying national petroleum needs, including increasing domestic crude oil production, developing a synthetic fuels industry, establishing commercial treaties with petroleum-exporting states, and occupying Romania, a major European oil producer, in a "bloodless invasion."[85] Like Japan's aggression in Southeast Asia, Germany's oil campaign occurred in the midst of an ongoing conflict: in this case, World War II in Europe (1939–1945). The conflict, and the territorial expansion that preceded it, increased Germany's oil consumption and limited its access to foreign resources. However, the European war was not caused by petroleum ambitions. Instead, it emerged from Hitler's desire to make Germany the dominant power in the region, if not the world.

Hitler's hegemonic aspirations arose partly from the outcome of World War I. Germans strongly resented the punitive Treaty of Versailles (1919), which prevented their country from remilitarizing the Rhineland and compelled it to pay harsh reparations to the war's victors. These obligations impeded Germany's postwar recovery and prevented the state from attaining the European leadership role that many of its citizens believed it deserved.[86] Germany, like Japan, also suffered intensely during the Great Depression. The value of its trade fell more significantly than the other great powers', and its drop in exports, coupled with reparations payments, liquidated most of Germany's foreign exchange reserves.

Consequently, the state could not pay for the foreign agricultural products and raw materials it needed to feed its people and power its industrialized economy.[87] Germans, including Hitler, increasingly lost faith in the global trading system. As the leader stated in 1934, "Our dependence on foreign trade would condemn us eternally to the position of a politically dependent nation."[88]

To address these concerns, Hitler invoked the concept of lebensraum. As observed in chapter 2, the concept was not new. Geopolitical thinkers introduced it in the 1890s when applying Darwinian logic to nation-states. The inventor of the term, Friedrich Ratzel, linked lebensraum to settler colonization; he believed that the populations of crowded states must move into newly conquered territories, where they would engage in small-scale agriculture to develop connections to their new lands. Hitler's vision of lebensraum shared this agrarian orientation. In *Mein Kampf*, he asserted that Germany's economy should revolve around farming, with industry and commerce playing secondary roles. An inveterate Malthusian, Hitler was also preoccupied with Germany's inadequate domestic food supplies. He believed that territorial expansion, particularly into eastern Europe, was the only way to rectify these shortages.[89] As historian Ian Kershaw writes, "Only through expansion—itself impossible without war—could Germany, and the National Socialist regime survive. This was Hitler's thinking. The gamble for expansion was inescapable. It was not a personal choice."[90]

This perceived need for expansion, coupled with Hitler's desire to dominate Europe, drove his initial acts of aggression. In March 1935, Hitler announced plans for national rearmament. A year later, he remilitarized the Rhineland, adjacent to France. In March 1938, Germany annexed Austria (the Anschluss), and in September 1938, German forces occupied the Sudetenland, a province of Czechoslovakia. The next March, Germany violated the terms of the Munich Agreement by seizing the Czech provinces of Bohemia and Moravia and, on September 1, initiated World War II by invading Poland. By June 1940, Hitler had conquered Denmark, Norway, Belgium, the Netherlands, Luxembourg, and France. The Battle of Britain began in July 1940.

The opening year of World War II, as well as the territorial expansionism that preceded it, was not entirely divorced from natural resource concerns. By acquiring more territory, Germany could expand the amount of arable land and raw materials under its direct control. The annexations of Austria and Czechoslovakia, in particular, were undertaken in light of Hitler's lebensraum aims. However, oil acquisition, specifically, was not a significant German goal. The Hossbach Memorandum, which describes a November 1937 meeting where Hitler emphasized his raw material concerns, does not mention oil.[91] In addition, none of Germany's early territorial acquisitions substantially increased the country's petroleum endowments. When the Anschluss occurred, Austria's oil industry was still in its infancy. The state's first commercial discovery occurred in 1934 and, in 1938, its output was only 384,000 barrels per year, making the state a net oil importer.[92] The Czech province of Moravia, adjacent to Austria's oil zone, also produced only nominal amounts of petroleum; the state's total annual output, including the Slovak provinces that Germany did not initially seize, was only 210,000 barrels.[93]

Poland was a more significant oil producer. However, when Germany and the USSR divided the conquered country into separate spheres of influence in September 1939, Hitler did not insist on gaining full control over Galicia, Poland's primary oil-producing region. Instead, the Soviets acquired control over 70 percent of Polish petroleum reserves.[94] Germany also received little oil from seizing France, as the state's fields at Pechelbronn, in Alsace, produced only 420,000 to 450,000 barrels of oil per year.[95] The Germans did seize large stocks of petroleum products when they conquered France and the Low Countries: almost 6.4 million barrels of aviation gasoline, motor gasoline, diesel oil, and fuel oil, by one estimate. However, these stocks would fulfill Germany's petroleum needs for less than two months.[96]

Instead of significantly increasing Germany's oil endowments, the state's territorial expansion from 1938 to 1940 decreased its energy security. Since the mid-1930s, Germany's oil demand had skyrocketed, as a result of the country's recovery from the Great Depression and its intense remilitarization program. Between 1936 and 1938 alone, German oil consumption jumped from 90,000 barrels per day to 150,000 barrels per day.[97] After the state's victories in central and western Europe, Germany was also responsible for much of the continent's petroleum needs, which had to be met to maintain the territories' economic productivity. To avoid a drastic shortfall in supplies, German authorities implemented draconian consumption cuts in occupied Europe and at home. Nevertheless, entering 1941, Axis Europe faced an annual petroleum shortage of at least 125-million-barrels.[98]

This shortfall arose because, in addition to increasing Germany's oil demand, the state's aggression provoked international retaliation, which undermined its access to foreign petroleum supplies. In March 1936, in response to Germany's remilitarization, Romania increased its oil prices and the USSR temporarily halted petroleum exports to Germany. More significantly, following Germany's invasion of Poland, the United Kingdom and France blockaded the country, cutting off seaborne shipments of crude oil and petroleum products.[99] Before the blockade, over 70 percent of Germany's oil imports had arrived from overseas: in particular, from the United States and Dutch West Indies (specifically, Aruba and Curaçao, which refined Venezuelan oil).[100] With these supply lines severed, Germany's oil crisis intensified.

Nonetheless, until summer 1941, Hitler refrained from launching an oil campaign, instead relying on a variety of peaceful strategies to satisfy national petroleum needs. One of these was increasing domestic oil production. From 1933 to 1937, exploratory drilling in Germany tripled, as a result of government subsidies and import duties on foreign gasoline. This expanded exploration did not result in significant new domestic discoveries, although output from existing oil fields rose from 1.6 million barrels in 1932 to 7 million barrels in 1940. However, this production bump was neither very useful nor sustainable. German crude oil was not suitable for producing aviation fuel, which was vital to the state's war effort, and output from domestic fields declined after 1940.[101]

Germany had far more success developing synthetic fuels. By 1927, the I. G. Farben company was using the Bergius hydrogenation process to produce aviation gasoline, motor fuels, and diesel fuels from Germany's abundant coal resources. After Hitler became chancellor in January 1933, the government significantly increased its support for the program, guaranteeing markets and prices for specified levels of fuel output.[102] The synthetic fuels program was also a central component of Germany's Four-Year Plan, initiated in October 1936, after Hitler's "Confidential Memo on Autarky" stressed the importance of German self-sufficiency in critical raw material inputs. Hitler recognized that complete autarky was impossible, particularly for foodstuffs. Nonetheless, he wanted Germany to be self-sufficient in fuel supplies within eighteen months. Money was to be no object in the pursuit of synthetic fuels; as Hitler put it, "The question of the cost of these raw materials is . . . quite irrelevant."[103] To maximize production, the state also exerted increasing control over I. G. Farben's operations.[104]

In 1936, the German government announced plans for the construction of ten new synthetic fuel plants, with an estimated output of approximately eighteen million barrels per year by 1939. Although the state did not achieve this ambitious target, by the time Germany invaded Poland, fourteen hydrogenation and Fischer–Tropp plants were operating and six more were under construction. That year, existing plants produced over ten million barrels of fuel.[105] By 1940, synthetic fuels provided up to 46 percent of Germany's oil supply and 95 percent of the Luftwaffe's fuel.[106] This output was extraordinary, particularly in comparison with the dismal performance of Japan's synthetic fuels program. However, output still fell short of production targets, partly as a result of shortages of steel and labor for plant construction.[107] Consequently, domestic resource development, alone, failed to satisfy Germany's oil needs.

Recognizing these limitations, Germany turned to foreign trade to acquire additional petroleum resources. The state's commercial initiatives targeted Romania and the USSR, Europe's two leading oil producers. The former's output, in 1937, was approximately sixty-one million barrels of crude and refined products, making it a particularly appealing trading partner.[108] However, Romania historically exported little oil to Germany, partly because of western European companies' domination of the local industry. When Berlin intensified its efforts to obtain a larger share of the state's oil sales in 1938, the Romanian government, wary of increasing its economic ties to Germany, approached British and French authorities, trying to persuade them to counterbalance the growing pressure. However, Bucharest was operating under duress, particularly after the Anschluss and Germany's occupation of the Sudetenland. King Carol feared that his country was Germany's next target. Accordingly, when he met with Hitler in November 1938, the ruler was willing to make significant concessions in order to avoid an invasion or complete economic domination. On March 23, 1939, a week after German forces seized Moravia and Bohemia, Berlin and Bucharest established a commercial agreement that included provisions for greater German involvement in the Romanian oil industry.[109]

Following the agreement, Romania's efforts to engage the British and French finally bore fruit. In April 1939, Britain sent an economic mission, led by Frederick Leith-Ross, to strengthen commercial ties with Romania and promote Western investment in the country. The United Kingdom and France also signed economic agreements with Bucharest and, on April 13, guaranteed Romania's territorial integrity. When the war began six months later, the British government authorized purchases of Romanian petroleum to keep it out of German hands. Western oil companies' Romanian subsidiaries also declined to sell resources to Germany and impeded petroleum transportation by refusing to lease rail tanker cars and chartering all available river barges. The Allies even tried to pay the Romanians to destroy their oil fields in the event of a German invasion; however, they failed to agree on a price.[110]

These Western interventions nonetheless failed to assuage Romanian fears, particularly since the Anglo–French territorial guarantee applied only to a German attack, not a Soviet one, which the Romanians regarded as equally threatening. Germany and the USSR's subsequent establishment of the Molotov–Ribbentrop Pact in August 1939, followed by their joint invasion of Poland, persuaded Romanian authorities that they needed stronger protection against the Soviets, which only Berlin could provide. Romania also desperately needed armaments, which the Germans were willing to supply in exchange for oil. Negotiations on a new economic agreement began by December 1939, and the states reached a preliminary accord on March 7, 1940. On May 27, following Hitler's rapid advance through France, the states signed an oil pact, marking Romania's "decisive shift towards Germany."[111]

At the same time, the Romanian government was strengthening its authority over the petroleum industry by establishing new regulations and imposing controls over movement in the oil regions around Ploiești. Germany also acquired direct control over portions of the industry, as a result of its victories in western

Europe. By defeating France, the Netherlands, and Belgium, the Germans gained control over multiple Western oil companies, including their Romanian subsidiaries. In July, Berlin installed a pro-German director for Astra Romana, Royal Dutch Shell's Romanian subsidiary, which controlled more than a quarter of the country's oil. In August, German "controllers" took charge of other oil companies. The same month, Western companies' foreign senior management and technical personnel were forced out of Romania.[112]

This degree of control might have satisfied German petroleum concerns, had it not been for subsequent Soviet actions. On June 22, 1940, Moscow demanded that Romania cede the territories of Bessarabia and Bukovina: two areas Bucharest had seized from the USSR at the end of World War I. The ultimatum antagonized Germany; the Molotov–Ribbentrop Pact had included Bessarabia in the Soviet sphere of influence, but did not permit occupation of the territory or mention Bukovina.[113] Nonetheless, Berlin expressed little overt opposition, and the beleaguered Romanians relinquished the territories. Fearing further attacks from Hungary or Bulgaria, Bucharest subsequently moved even further into Berlin's orbit. On June 29, the government renounced the Anglo–French territorial guarantee, and the next month, Germany stationed bomber units in Braşov, Transylvania, in central Romania.[114]

Hitler had little interest in preserving Romania's territorial integrity; he was only interested in securing the state's petroleum industry and consolidating Germany's position in eastern Europe. To achieve these ends, Berlin pushed Bucharest to cede additional territory to its neighbors. In the Second Vienna Award (August 30, 1940), Romania surrendered northern Transylvania to Hungary, and in the Treaty of Craiova (September 7, 1940), it yielded southern Dobruja to Bulgaria. Romanian citizens were infuriated by these territorial losses and, on September 6, compelled King Carol to abdicate. He was succeeded by General Ion Antonescu, a German sympathizer, who imposed stronger state controls over the economy and aligned the country more firmly with Germany, which guaranteed Romania's new borders. By October, more German forces had arrived in the country, partly to guard Romania's oil fields. On November 23, Romania joined the Tripartite Pact. Less than two weeks later, the countries signed a new economic agreement, which extended the May accord and effectively "reorganize[d] the entire economy . . . under German auspices." From that point on, while Bucharest formally retained authority over its petroleum resources, in practice, Berlin possessed full control.[115] Germany had acquired its neighbor's oil without firing a shot.[116]

Germany's petroleum purchases from the USSR were significantly smaller than its imports from Romania. Although the Soviets produced far more oil—an estimated 203 million barrels per year in the late 1930s—they consumed most of their

MAP 7.2. Eastern Europe and the Caucasus (December 1940)

output domestically, leaving little surplus for export.[117] Nevertheless, the USSR was an appealing German supplier, because of the complementarity of the states' economies. In the 1930s, Germany needed food and raw materials, which the Soviets possessed in abundance. The USSR, in contrast, required German manufactured goods, especially industrial machinery and armaments. As a result, the Soviets were willing to establish barter-based commercial agreements, obviating Germany's foreign exchange problem. The states reached minor economic agreements in spring 1935 and 1936. However, over the next two years, they failed to accomplish more than renew the accords, because of purges in the Soviet government and Hitler's reluctance to grant any political concessions to the USSR.[118]

By late 1938, however, resource imperatives pushed Germany back to the negotiating table. The prospect of a major war over Poland, which arose when the United Kingdom and France guaranteed the state's territorial integrity in March 1939, also amplified incentives to create a commercial accord. On August 20, 1939, Germany and the USSR finally signed a "credit agreement," which would enable the former to acquire vital raw materials and the latter to obtain industrial goods. Following their joint invasion of Poland, the states also signed the Boundary and Friendship Treaty (September 28, 1939), in which the USSR agreed to supply Germany with all output from the Soviet-occupied Galician oil fields in return for steel tubing and coal resources. Germany therefore gained access to all of Poland's oil without conquering it directly. On February 11, 1940, Berlin and Moscow established a new agreement promising larger volumes of trade, including deliveries of 6.3 million barrels of Soviet oil. They signed contracts for oil prices and delivery schedules in late May.[119]

Hitler did not trust commercial agreements; he believed that they "afforded no guarantee for actual execution."[120] In the case of the USSR, these concerns proved well founded. Soviet raw material deliveries persistently failed to match promised levels; in January 1940, for example, oil shipments fell 240,000 barrels short. By May 1940, the USSR had delivered only 1.1 million barrels of petroleum, despite agreeing, the previous November, to provide 1.4 million barrels of fuel by January. By August 1940, the Soviets had delivered only 31 percent of the raw materials pledged in the February agreement. In September 1940, Moscow threatened to halt resource shipments entirely if the Germans failed to deliver manufactured goods on schedule. The risk of deliberate supply shutoffs was therefore a persistent concern for Berlin. In addition, officials recognized that, over time, the Soviets' ability to export oil would decline, as a result of their rising domestic petroleum consumption.[121] Nonetheless, the states signed another economic agreement on January 10, 1941.

Hitler, however, had already decided to invade the USSR. Oil concerns contributed to this choice, as the state had proved to be an unreliable trading partner.[122] Romanian oil output was also declining, from sixty-one million barrels in 1937 to thirty-nine million barrels in 1941, suggesting that Germany would soon have to look elsewhere for supplies.[123] The Western Hemisphere's petroleum resources remained inaccessible and Germany's own crude output had peaked. Meanwhile, the state's synthetic fuels program had failed to create self-sufficiency. As Hitler observed in June 1941, "The course of the war shows that we have gone too far in our efforts to achieve autarky. It is impossible to produce all that we lack by synthetic processes."[124] In addition, Hitler was very concerned about the Soviet threat to Romania's oil industry. The USSR's occupation of Bessarabia, following its June 1940 ultimatum, had brought Soviet forces within 120 miles of the Ploieşti oil fields: far too close for German comfort.[125] Noting the Soviets' menacing presence, Hermann Göring, director of the Four-Year Plan, observed, "Perhaps we shall be forced to take steps against all this, despite everything, and drive this Asiatic spirit back out of Europe and into Asia, where it belongs." Hitler, speaking to Mussolini, was more concise: "The life of the Axis depends on those oilfields," he asserted.[126]

Operation Barbarossa, Germany's invasion of the USSR, can therefore be labeled an oil campaign. Directive 21 (December 18, 1940), which outlined the original invasion plan, did not mention any oil-bearing regions by name. However, it asserted that German forces should advance to the Volga–Archangel line, stretching from the city of Astrakhan, on the Caspian Sea, to the city of Archangel (Arkhangelsk), on the White Sea. Conquered territories would therefore include the Caucasus region, which produced 90 percent of the USSR's oil. Hitler also explicitly identified Baku, the source of 80 percent of the USSR's oil output, as a target during a war conference at the Berghof on July 31, 1940.[127] In early 1941, German economic planners, including the head of the War Economy and Armaments Office, General Georg Thomas, also highlighted the importance of the Caucasus's oil, both for powering Germany's war effort and for sustaining the productivity of agriculture and industry in occupied Soviet territories.[128] Seizing Ukraine's rich agricultural lands—another central target of Operation Barbarossa—would be useless, unless Germany also obtained the Caucasus's oil.[129]

Yet petroleum ambitions were far from the only reason for Germany's invasion. Hitler had been anticipating a "war of destruction" with the USSR since the mid-1920s.[130] He abhorred Bolshevism, which he associated with Judaism, and viewed its eradication as his "life mission." By 1936, Hitler had come to believe that this "historical conflict" must occur soon, before the Soviets' continuing military and economic development rendered a German victory impossible.[131] Conquering the USSR would also achieve Hitler's long-standing aspiration for lebensraum by providing Germans with abundant arable land and raw materials.[132] In addition, the German leader believed that defeating the Soviets would force the United Kingdom out of the war. As the Battle of Britain dragged on, Hitler was increasingly convinced that London was refusing to sue for peace because it thought that the USSR would intervene on its behalf. However, "with Russia smashed," Hitler surmised, "Britain's last hope would be shattered." Defeating the United Kingdom would also ensure that the United States did not enter the war.[133]

Operation Barbarossa began on June 22, 1941, with a three-pronged attack. Army Group North advanced through the Baltics toward Leningrad; Army Group Center moved toward Moscow, targeting the USSR's political leadership; and Army Group South advanced through the Lublin area of Poland, then attacked Soviet forces west of the Dnieper River and proceeded toward Kiev, with the subsequent aim of capturing the industrial Donets Basin.[134] Initially, all three groups advanced swiftly. Directive 32a (July 14, 1941) was the first to mention oil. Discussing postwar planning for a conquered USSR, it noted that it would be "particularly important to ensure supplies of raw materials and mineral oil." Directive 33a (July 23, 1941) finally identified an oil-rich target by name. The directive instructed Army Group South to defeat Soviet forces west of the Dnieper, occupy the Donets industrial area, then cross the Don River and advance toward the Caucasus.[135]

Before advancing into the Caucasus, however, Germany prioritized oil defense and denial. Directive 34a (August 12, 1941) instructed German forces to "occupy the Crimean Peninsula, which is particularly dangerous as an enemy air base against the Rumanian oil fields."[136] As Hitler expected, the USSR had initiated air attacks on Ploiești a few days after Operation Barbarossa began. By mid-July, these assaults had caused significant damage and, since Romania supplied the majority of German oil imports, they were a serious threat to the state's energy security. As Hitler observed on August 22, "It is of decisive importance for Germany that the Russian air bases on the Black Sea be eliminated. . . . This measure can be said to be absolutely essential. . . . Such attacks could have incalculable results for the future conduct of the war."[137] The German leader also aspired to cut Soviet supply lines from the Caucasus to the rest of the USSR. By denying the Soviets fuel, Germany could impair its adversary's military operations and industrial production, which ran predominantly on oil. Accordingly, on August 21, Hitler claimed that "the most important aim to be achieved before the onset of winter . . . is not to capture Moscow, but to seize the Crimea and the industrial and coal region on the Donets, and to cut off the Russian oil supply from the Caucasus area."[138]

By November 1941, it was clear that Operation Barbarossa would meet none of these oil-related objectives. German forces failed to seize the entirety of Crimea. They did not sever Soviet supply lines. And Army Group South advanced only as far as Rostov, on Ukraine's eastern border—over 150 miles from the nearest Caucasian oil fields, at Maikop—before a Soviet counteroffensive forced it to withdraw from the city in December. Meanwhile, Army Group Central's advance ground to a halt twenty miles outside Moscow.

In addition to failing to defeat the USSR, Operation Barbarossa left Germany distressingly short of oil. At the beginning of 1942, German fuel stockpiles could supply no more than two months of civilian and military consumption.[139] In February, the War Economy and Armaments Office warned that, without more oil, Germany could not power its war machine or exploit the Soviet areas it already occupied. It would certainly be unable to fuel the long struggle with the USSR, United Kingdom, and United States that Hitler now believed was inevitable.[140] The same month, General Antonescu told German foreign minister Joachim von Ribbentrop that, "as for crude oil, Rumania has contributed the maximum which it is in her power to contribute. She can give no more." "The only way out of the situation," the Romanian leader observed, "would be to seize territories rich in oil."[141]

German officials concurred. On April 5, 1942, they issued Directive 41, which outlined Case Blue, a new offensive against the USSR. In contrast to Operation Barbarossa, Case Blue was driven almost entirely by petroleum ambitions. The campaign prioritized seizing the Caucasus's oil fields: at a minimum, those at Maikop and Grozny, which produced 10 percent of the USSR's petroleum, but ideally also the massive reservoirs at Baku. Recognizing the resources' significance, Hitler told senior officers on June 1, 1942, "If I do not get the oil of Maikop and Grozny then I must end this war."[142] Like Operation Barbarossa, Case Blue also aimed to conquer Crimea, eliminating the Soviet threat to Romanian oil fields, and sever the USSR's petroleum supply lines in order to undermine its warfighting capabilities. As Ribbentrop predicted to his Italian counterpart, Galeazzo Ciano, in April, "When Russia's sources of oil are exhausted she will be brought to her knees."[143] Still, petroleum was not the only target of Germany's 1942 campaign. In addition to reiterating the "decisive importance of the Caucasus oilfields for the further prosecution of the war," Directive 45 (July 23, 1942) split the southern army group in two. Army Group A would advance toward the Caucasus in Operation Edelweiss. Army Group B would conquer Stalingrad in Operation Heron.[144]

German officials were well aware of the obstacles to oil conquest before launching Operation Barbarossa and Case Blue. Invasion impediments would be severe, as retreating Soviet forces were expected to sabotage local oil installations. This would impair Germany's ability to refine Soviet crude oil locally for use on the Eastern Front, compelling the German army to transport captured crude to central Europe. A March 1941 report by the War Economy and Armaments Office highlighted the difficulty of shipping Soviet oil along the 2,500-mile route from the Caucasus to central Europe, especially with an already limited supply of railway tanker cars and river barges. In addition, capturing local fuel would not benefit advancing German forces, as the material's low octane content made it unsuitable for their vehicles. The Germany army could consume the fuel after enhancing it with benzol additives; however, this process required specialized facilities, which were not available locally. Recognizing these impediments, some of Hitler's advisers questioned the economic rationale for the Soviet campaigns.[145] To moderate anticipated invasion obstacles, Germany attached special "oil commandos" to Army Group South and created a Technical Oil Brigade to restore damaged petroleum installations.[146]

On May 8, 1942, Germany launched Case Blue by attacking the Crimean Peninsula. German troops captured the Kerch Peninsula, in eastern Crimea, by the end of the month and, by July 4, seized Sevastopol, home to the USSR's main Black Sea naval base.[147] Case Blue's main offensive began on June 28 and, like Operation Barbarossa, initially proceeded swiftly. Army Group A recaptured Rostov on July 28 and reached Maikop on August 9. There, however, they met with disappointment. As German officials had feared, retreating Soviet forces had perpetrated a comprehensive destruction campaign, cementing and setting fire to most of the area's oil wells and destroying a critical oil refinery at Krasnodar. In September, the Technical Oil Brigade reported that it would take at least six months to restore Maikop's regular production.[148]

Even that assessment soon proved to be overly optimistic. The initial damage was compounded by the difficulty of obtaining drilling equipment to repair oil wells and by occupation obstacles in the form of ongoing attacks by local guerillas. As a result of these impediments, German engineers were unable to extract any oil from Maikop until December.[149] Meanwhile, Army Group A's advance slowed as a result of Hitler redirecting forces from the Caucasus to Stalingrad, intense Soviet resistance, and—ironically—inadequate fuel supplies. In November, Group A was halted fifty-five miles outside Grozny. The next month, a Soviet counteroffensive pushed the Germans back to Maikop and, in January 1943, out of the Caucasus entirely.[150] German forces would never return to the region; Hitler's oil campaigns had failed.

Before launching his oil campaigns, Hitler was less perturbed by the petroleum-related obstacles to international aggression than Japanese leaders were. Although German officials acknowledged that a drive to the Caucasus would entail significant invasion impediments, they expressed less concern about occupation or international obstacles—perhaps because they had already obviated the latter by defeating most of their adversaries. Nevertheless, Hitler's reliance on alternative strategies to satisfy national oil needs, for more than five years before attacking the USSR, suggests that the German leader also viewed foreign aggression as an unappealing means of acquiring petroleum resources. Hitler did not fight for oil until he had exhausted every other option. Moreover, while Operational Barbarossa and Case Blue targeted oil, obtaining petroleum resources was not the German offensives' only goal.

The German and Japanese oil campaigns reveal that petroleum concerns can influence wars' trajectories once they are under way. Both aggressors launched attacks that largely aimed to grab foreign oil resources. However, the energy insecurity that motivated these campaigns emerged from the states' existing conflicts in Europe and East Asia, which were caused by their desire for regional hegemony, not oil. These ongoing conflicts heightened both countries' petroleum needs, impaired their ability to satisfy them peacefully, and created permissive conditions for further international aggression. In the absence of these wartime conditions, however, Japan and Germany would have refrained from seizing foreign petroleum resources.

8

OIL GAMBIT

Iraq's Invasion of Kuwait

Iraq's invasion of Kuwait in 1990 is often regarded as the quintessential classic oil war: the country launched a major conflict aimed at grabbing its neighbor's petroleum resources in the absence of an ongoing war. Even oil war skeptics regularly identify this case as an exception to their argument that states avoid fighting for petroleum resources.[1] Some classic oil war interpretations of the conflict emphasize Iraq's purported oil greed, asserting that Saddam Hussein invaded Kuwait because he believed that aggression would be profitable.[2] By seizing his neighbor's resources, Iraq would reap enormous wealth and become the world's dominant petroleum producer.[3] As evidence of the state's oil ambitions, these "greedy" narratives also emphasize Baghdad's complaint, before the invasion, that Kuwait was slant drilling into the transboundary Rumaila oil field to "steal" Iraqi resources. As the Duelfer Report, on Iraq's pre-2003 weapons of mass destruction (WMD) programs, claimed, "The impulsive decision to invade in August 1990 was precipitated by . . . negotiations over disputed oil drilling along the common border."[4]

Other classic oil war interpretations argue that Saddam was acting out of oil need, not oil greed.[5] In the months before the attack, they observe, Iraq faced an oil-related economic crisis. The state had emerged from its war with Iran with expansive debts and needed abundant petroleum revenue to repay them. Yet oil prices were declining, largely as a result of other Persian Gulf producers—particularly Kuwait and the United Arab Emirates (UAE)—exceeding their OPEC output quotas. Iraqi leaders feared that, if the price of oil did not rise, they would no longer be able to finance domestic social spending, which would threaten their regime's security. Eventually, Saddam concluded that seizing Kuwait's oil fields offered the only possible means of alleviating his state's economic and political crisis.

Yet even this "needy" oil war interpretation is an oversimplification. Iraq's economic crisis, alone, did not drive its international aggression. Instead, the economic threat was magnified by Saddam's conviction that the United States was driving Kuwait's and the UAE's actions. He reached this conclusion because, like Japanese authorities before World War II, he believed that the US government was determined to resist his country's regional rise. If manipulating international petroleum output and prices failed to quash Iraq and overturn its Baʿathist regime, the United States would resort to other, more aggressive tactics, most likely with the assistance of its regional ally, Israel. It was this belief, that Iraq faced a broader existential threat, that precipitated Saddam's attack. This conviction also enabled Saddam to assert, after the conflict, that he had won the war.[6] Although Iraq had obtained no petroleum payoffs—and was actually worse off economically than before the conflict, as a result of international sanctions—the state had confronted the world's sole superpower in the "mother of all battles" and survived.[7]

Even with this broader incentive for aggression, however, Saddam approached his oil gambit circumspectly. Like the perpetrators of oil spats, the Iraqis launched their attack in a dispute scenario. From Baghdad's perspective, Iraq and Kuwait were engaged in a long-standing territorial contest, which involved, at a minimum, the islands of Warba and Bubiyan, at the eastern terminus of the states' land boundary, and, at a maximum, the entirety of Kuwait. Because of this ongoing territorial disagreement, Iraq's aggression had a patina of legitimacy for some international observers. The dispute also meant that, if Iraq acquired Kuwait, it would gain more than oil resources. The state would also enhance its economic and military security by improving its sea access and strengthen the regime's domestic standing by defeating a longtime opponent.

In addition, like the aggressors in oil campaigns, Iraq did not rush to seize foreign petroleum resources. Instead, Saddam's invasion of Kuwait was an "act of last resort."[8] Contrary to the conventional wisdom, the Iraqi leader did not attack because he believed that he had received a "green light" from US ambassador April Glaspie. Instead, Saddam recognized that the United States would retaliate for his aggression, economically and militarily. To forestall that response, Iraqi authorities initially attempted to manage their crisis in other ways. They initiated domestic economic reforms. They repeatedly approached other Gulf oil producers, asking them to rein in their petroleum output and cancel Iraq's war debts. They also sought reassurance from American officials that the United States did not harbor hostile intentions toward Baghdad. Saddam refrained from launching his invasion until all of these alternative, peaceful initiatives had failed.

Iraq's oil gambit was therefore initiated instrumentally, selectively, and reluctantly. These characteristics suggest that, if the Iraqi invasion offers the strongest historical evidence of classic oil wars, believers are on shaky ground. Decision makers' willingness to fight for oil, even in this most likely case, was highly circumscribed. Hence, if we choose to call oil gambits classic oil wars, we must recognize that these conflicts look quite different from the greedy petroleum grabs that we usually imagine. Fighting for oil is not an appealing prospect, even for the world's most ruthless leaders. Nor is it an appealing prospect for unrivalled superpowers; as I will argue in a postscript to this chapter, the United States' subsequent invasion of Iraq in 2003 was not a classic oil war.

Fighting for Survival

Unlike oil campaigns, Iraq's oil gambit did not occur in the midst of an ongoing war; the state's conflict with Iran had ended almost two years earlier. Nonetheless, the Iran–Iraq War set the stage for Saddam's attack; indeed, it is difficult to imagine Iraq invading Kuwait without it. To sustain the earlier conflict, Iraq had borrowed extensively from Arab and Western creditors. By the time Tehran agreed to a ceasefire in August 1988, Baghdad owed over $80 billion.[9] Servicing these war debts constituted a major financial burden for the Iraqi government, consuming at least $5 billion annually. Meanwhile, the country's annual gross domestic product (GDP) was only $25 billion.[10] In addition, Iraq needed to finance domestic reconstruction, which was expected to cost $230 billion, and maintain social spending for a population that was exhausted by war and eager for a return to normalcy.[11] Saddam himself heightened Iraqis' peacetime expectations by trumpeting the state's supposed "victory" in the Iran–Iraq War.[12] The regime therefore faced a mounting domestic crisis. Officials feared that if they failed to improve economic conditions, Iraqis' already degraded standard of living would continue to fall, intensifying hostility toward the government and eventually leading to a domestic uprising.[13]

To a large extent, Iraq's trajectory is therefore consistent with Germany's and Japan's before World War II. All three states engaged in initial acts of aggression, which heightened their petroleum needs, provoking economic and political crises. Eventually, each state concluded that further foreign conquest offered the only means of fulfilling its expanded oil requirements. One difference between these cases, however, was the character of aggressors' petroleum needs. Japan and Germany were oil consumers, so their initial attacks increased national demand for oil *resources*; the states required more petroleum to sustain their ongoing wars in China and Europe. Iraq, in contrast, was an oil producer, deriving 60 percent of its GDP and 95 percent of its foreign currency earnings from petroleum sales.[14] Consequently, Iraq's initial act of aggression—its war with Iran—increased Baghdad's need for oil *revenue*. After the conflict, the Iraqi government required high oil prices to service its debts and cover domestic expenses.

To officials' great consternation, however, international oil prices were falling. From January to July 1990, they dropped from $20.50 per barrel to $13 per barrel.[15] Iraq's budget was based on an $18 per barrel price, so the collapse constituted a major financial burden.[16] The principal cause of the price drop was Kuwait and the UAE exceeding their OPEC production quotas. The two countries had increased their oil output during the Iran–Iraq War to compensate for the belligerents' lowered production. When the conflict ended, Kuwait and the UAE resisted reducing their output to accommodate the Iranian and Iraqi oil supplies that were returning to the market. In 1989, Kuwait's output exceeded its OPEC quota by approximately seven hundred thousand barrels per day.[17] Because of this overproduction, Iraq could not pump its way out of its economic crisis; any increase in oil output would trigger a further drop in oil prices.[18]

Instead, the Iraqi government attempted to improve its economic situation through domestic reforms. These were initiated in the early years of the Iran–Iraq War and expanded after the ceasefire. They included a massive privatization initiative, which transferred most land and agricultural production from state to private hands. The government also sold many state-owned industries and opened the economy to increased foreign investment, with the aim of expanding nonpetroleum exports. Economic productivity initially rose, as a result of major cuts in the workforce. However, only a small proportion of the population benefited from the reforms. Income inequality, which was previously minimal in Iraq, increased dramatically, along with unemployment. The removal of price controls on many goods, including foodstuffs, triggered inflation. In addition, cutbacks on state support for imports precipitated shortages of basic goods.[19] Meanwhile, international creditors, concerned about the government's lack of financial transparency and ability to service its debts, began to cut back on their lending. By the late 1980s, Iraq was worse off economically than it had been before the reforms. Domestic discontent intensified and threatened a collapse in civil order.[20]

The Iraqi government could have improved its financial position by reining in military spending. After the ceasefire, Baghdad demobilized only a small portion of its armed forces and continued to invest heavily in weapons research and development, so military expenditures remained high. However, in the Iraqis' view, domestic and international security concerns precluded larger cuts. Broader demobilization would intensify domestic instability, as Iraq's tattered economy could absorb no further increases in the workforce and discontented ex-soldiers would pose a serious threat to the regime.[21] Internationally, the Iraqi government continued to view Iran as a significant security threat. The Islamic Republic's capabilities had been degraded by the war but not eliminated, and Iran possessed an inescapable demographic advantage over Iraq. The Iranian government also refused to implement the states' August 1988 ceasefire or negotiate a permanent peace treaty. Consequently, Iraq was determined to maintain a military advantage over its former adversary to discourage future aggression. In addition, Iraqi officials believed that they needed to deter attacks from their other long-standing regional adversary: Israel.[22]

Baghdad therefore faced a major economic crisis related to oil revenue, which threatened to escalate into a political emergency. However, these dire conditions, alone, failed to provoke international aggression.[23] Instead, it was Saddam's belief that the United States was driving the crisis that elevated these conditions to a perceived existential threat. In the two years between the end of the Iran–Iraq War and his invasion of Kuwait, Saddam became convinced that the United States was irremediably hostile to his regime. He assumed that the US government was driving Kuwait's and the UAE's overproduction, interpreting it as the Americans' latest tactic in a long-standing plan to resist Iraq's regional rise and remove him from power.[24] If manipulation of Iraq's oil revenue failed to unseat him, Saddam believed, the United States would eventually turn to assassination attempts, airstrikes, missile strikes, or an occupation to overturn his regime and undermine the country.[25] As Saddam put it later, "They wanted to force our status backwards . . . to crush us spiritually and force us to abandon our role."[26]

Saddam's perception of the United States' implacable hostility was grounded in historical experience. In public and private statements, the Iraqi leader repeatedly referred to the United States as an "imperialist" power that aimed to maintain a hegemonic role in the Middle East. He believed that American aspirations in this pursuit were twofold; their core interest was the region's oil and their secondary goal was to support their ally, Israel. To advance this "imperialist–Zionist" agenda, Iraqis surmised, the United States would oppose any local state that threatened its access to oil resources or challenged Israel.[27] Accordingly, Iraqi leaders assumed that US officials had been hostile to the Ba῾ath regime since it took power in 1963 because of the party's anti-imperialist stance. This antagonism had intensified, they believed, when Iraq fully nationalized its oil industry in 1972. In doing so, the state removed control of petroleum resources from Western hands and obtained expansive oil revenue to fuel its economic and military development, thereby increasing its regional power.[28]

Saddam also assumed that Iraq was the object of American ire because it was the "natural" leader of the Persian Gulf region. He based this grandiose assessment on the state's six-thousand-year cultural legacy, often comparing himself to historical leaders, such as Saladin.[29] In addition, after Egypt signed the Camp

David Accords in 1978, Saddam, like many other Arab leaders, believed that the state had forfeited its right to further regional leadership. Saddam asserted that Iraq, which had persistently championed the Palestinian cause, was Egypt's natural successor. Yet he also assumed that, by taking on the mantle of Arab leadership, Iraq would incur intensified "imperialist–Zionist" opposition.[30]

The first significant evidence of American hostility, Saddam believed, was the United States' support for the Kurdish rebellion against Baghdad from 1972 to 1975. The Nixon administration provided the Kurds with arms and financial assistance and facilitated equipment transfers from the Israeli government to Kurdish forces so they could sustain their military campaign. As discussed in chapter 5, Kurdish pressure eventually compelled Saddam to accept the Algiers Agreement with Iran. The Iraqi leader held Washington partly responsible for this "humiliating" outcome, as he was well aware of the United States' involvement in the conflict. In 1975, Saddam complained to an American delegation that "US strategy in the region was a pincer movement involving Israel and Iran directed at destroying the Iraqi revolution."[31] He interpreted the Rapid Deployment Joint Task Force's positioning in the Persian Gulf in 1980, as well as Israel's attack on Iraq's Osirak nuclear reactor in 1981, as further evidence of the "imperialist–Zionist" conspiracy against him.[32]

The episode that solidified Saddam's perceptions of the United States' unremitting hostility, however, was the Iran–Contra scandal. In the early stages of the Iran–Iraq War, the United States remained formally neutral and refused to sell weapons to Saddam's regime. However, it provided significant material support to Iraq, allowing it to purchase dual-use technologies, including trucks and helicopters. It also supplied Baghdad with signals intelligence and targeting data. Nonetheless, in a televised speech in November 1986, President Ronald Reagan revealed that the United States had been selling arms to Iran, first through the Israelis and then directly. Saddam was incensed by the apparent betrayal. "Irangate," as he referred to it, was a "stab in the back," which appeared to substantiate his belief that the United States was unreliable, perfidious, and determined to resist any increase in Iraq's regional authority. This conviction would color Saddam's subsequent interactions with his perceived adversary, encouraging him to interpret all apparent opposition in the worst possible light.[33] In his meeting with Ambassador Glaspie one week before the invasion of Kuwait, the Iraqi leader observed that "new events remind us that old mistakes were not just a matter of coincidence."[34]

After the ceasefire with Iran, Saddam anticipated that the United States' antagonism toward his regime would intensify. Although Iraq had been economically harmed by the war, it emerged from the conflict as the region's strongest military power. As a result, Saddam surmised, Iraq posed the greatest threat to the "imperialist–Zionist" agenda and would therefore be subject to more aggressive American containment efforts.[35] In addition, Saddam recognized that, following the collapse of the USSR, the United States would have a freer hand in the Gulf, as it would no longer be constrained by the threat of Soviet retaliation. As sole superpower, Washington could pursue its agenda against Iraq more vigorously.[36]

Iraqi–American interactions from 1988 to 1990 appeared to confirm Saddam's suspicions. After the Iran–Iraq War ended, the United States did not remove its military forces from the region. Instead, ships that had been deployed to the Gulf to protect Kuwaiti oil tankers remained in place. Saddam interpreted the open-ended presence as evidence of the United States' continued interest in maintaining its regional hegemony.[37] Around the same time, in autumn 1988, both branches of Congress passed bills condemning Iraq's use of chemical weapons against the Kurds and proposing economic sanctions.[38]

A year later, a scandal concerning the Italian Banca Nazionale Lavoro (BNL) threatened Iraq's access to US agricultural exports. Since 1983, the Department of Agriculture's Commodity Credit Corporation (CCC) program had been providing loan guarantees to banks that lent Iraq money to buy US agricultural products. By 1987, Baghdad was one of the program's largest customers. However, upon discovering that the BNL had diverted agricultural credits to facilitate Iraqi military purchases, multiple US government agencies pushed the White House to terminate the funds. Iraqi authorities perceived this initiative as a major security threat, as the state depended on US agricultural exports to feed its increasingly restive population. On October 6, 1989, in his first meeting with Secretary of State James Baker, Iraqi foreign minister Tariq ʿAziz protested the potential cutbacks, observing that they would endanger Iraq's ability to "feed its people" and attract further international lending.[39] Baker reported that ʿAziz also accused the United States of "interfering in Iraq's internal affairs and . . . conducting clandestine efforts to subvert their government."[40]

To reassure Baghdad, the White House and State Department obtained approval for $1 billion in CCC credits for Iraq in 1990. However, only half of that amount was immediately released; the second tranche was held in suspension, nurturing Iraqi suspicions of US intentions.[41] Baghdad's paranoia intensified in February 1990, when a Voice of America broadcast compared Saddam's regime to recently fallen dictatorships in Eastern Europe; the Iraqi leader interpreted the message as a direct threat. A week later, the State Department issued another report condemning Iraq's human rights record, and at the end of the month, after Iraqi missile launchers were discovered near the Jordanian border, Congress threatened to terminate all CCC lending. In March, US officials criticized Iraq for executing Farzad Bazoft, an Iranian-born journalist employed by the London *Observer*, whom the Iraqis accused of spying.[42]

Iraqi authorities believed that the United States was targeting their military capabilities, in particular, in order to sustain Israel's regional military superiority. In March, British scientist Gerald Bull, who was assisting Iraq in weapons development, was assassinated. The Iraqis blamed Mossad, Israel's national intelligence agency, which they assumed was acting with US assistance. Soon after, several European customs operations intercepted weapons materials, including possible nuclear triggers, bound for Iraq.[43] By spring, the regime had become convinced that another Israeli assault on its weapons or industrial facilities was imminent.[44] In early March, multiple senior Iraqi officials told the former assistant secretary of state Richard Murphy that they anticipated an Osirak-style attack. Later that month, Iraqi diplomat Nizar Hamdoon conveyed the same concern to the United Kingdom's chargé d'affaires.[45]

This conviction also motivated Saddam's infamous "burn half of Israel" speech at the beginning of April. The statement was intended as a deterrent; Saddam warned that he would retaliate against Israel *if* it attacked Iraq.[46] However, the White House immediately condemned the statement as "deplorable and irresponsible" and the Western media presented it as a signal of Saddam's aggressive intent. The Iraqis, in turn, interpreted this depiction as an American rhetorical campaign aimed at legitimizing an Israeli strike. The United States also responded to Iraq's apparent belligerence by continuing to block distribution of the second CCC tranche, and by May, members of Congress had introduced several bills calling for sanctions against Iraq.[47] 'Aziz reported later, "I was convinced . . . in April the Americans had stopped listening to us and had made up their minds to hit us."[48]

Faced with a burgeoning economic crisis and apparently unremitting US hostility, Iraqi leaders believed that they needed to take action to resist the intensifying existential threat. Invading Kuwait was one potential recourse. By occupying its neighbor, Iraq would gain control over additional oil resources, including the entirety of the giant transboundary Rumaila field.[49] Baghdad could therefore collect additional petroleum revenue, either by selling Kuwaiti resources or by enforcing production cutbacks, which would increase international oil prices and, consequently, the rents Iraq received from sales of its own and Kuwaiti oil. In addition, by controlling 20 percent of the world's petroleum reserves, Iraq would secure a dominant position in OPEC and global oil markets. Finally, by seizing Kuwait, Iraq would eliminate a portion of its war debts, obtain an outlet for its dissatisfied military forces, and acquire its neighbor's gold reserves and other valuable assets.[50]

Militarily, an invasion of Kuwait would be easy. Iraq's limited demobilization after the Iran–Iraq War, coupled with its continued military spending, meant that the state retained the strongest force in the region. Kuwait's terrain was difficult to defend and the country's military was weak. It would not be able to resist an

Iraqi attack or retaliate extensively against Iraq's oil infrastructure. Invasion obstacles were therefore low. The Iraqis also hoped that occupation obstacles would be limited, because of domestic political strife that destabilized the sheikdom from 1989 to 1990. "Disgruntled" Kuwaitis might be sympathetic to invaders that removed the ruling Sabah family.[51] Lastly, Saddam likely hoped that other Arab countries would refrain from retaliation. Within the region, Kuwaitis were widely regarded as arrogant, and other OPEC members, including Saudi Arabia, were also irked by their overproduction.[52] Hence, local states might be less inclined to come to Kuwait's aid, reducing international obstacles.

Iraq was also tempted to seize Kuwait because of the states' ongoing territorial dispute, which had begun in the early twentieth century. In 1913, the Ottoman Empire, which controlled the territories that would become Kuwait and Iraq, and the United Kingdom, which had established itself as Kuwait's protector in 1899, agreed to a convention identifying Kuwait as an autonomous *qaza* (district) within the Ottoman Empire. The outbreak of World War I prevented the accord's ratification. However, Kuwaitis believe that it established their state's independence. In contrast, Iraqis, whose country was created by merging the Ottoman *wilayat* (provinces) of Basra, Baghdad, and Mosul, argue that, as a *qaza*, Kuwait was never formally separated from Basra. Hence, the region remained part of Iraq: its "nineteenth province."[53] Since their own state's independence in 1932, Iraqis have periodically exploited this legal technicality to assert their authority over all or portions of Kuwait. From 1933 to 1939, Iraqi newspapers, likely with government support, published articles calling for the country's incorporation into Iraq.[54] In 1938, the Iraqi foreign minister, Taufiq as-Suwaidi, submitted an aide memoire to British diplomats, declaring Iraq's authority over the entirety of Kuwait.[55] The next year, Iraq's volatile King Ghazi publicly asserted that Kuwait was part of Iraq. In June 1961, a week before Kuwait achieved independence, Iraqi prime minister Abd al-Karim Qasim reiterated the claim.[56]

None of these initiatives were driven by oil ambitions. In 1961, for example, Qasim recognized that occupying Kuwait would not be worth the effort, for oil or any other purpose. As political journalist Peter Mansfield states, "Garrisoning a hostile Kuwaiti population supported by the rest of the Arab world would create endless problems for Iraq and Kassem [*sic*], although mercurial, could be rational in his strategic thinking."[57] Instead, Iraqi leaders used their territorial claim to improve their bargaining positions on other issues. In 1961, the fiercely anti-imperial Qasim was attempting to compel Kuwait to renounce its defense agreement with Britain in order to reduce Western influence in the region.[58] In the late 1930s, Iraq's central concerns were cross-border smuggling and access to the Persian Gulf.[59]

The Gulf access issue was the primary disagreement underpinning Iraq and Kuwait's enduring territorial dispute. As observed in chapter 5, Iraq's Persian Gulf coastline is only forty miles long. Since the 1930s, Iraqi leaders have persistently sought to annex or lease the islands of Warba and Bubiyan, at the eastern terminus of Iraq and Kuwait's land boundary, in order to enhance their country's meager sea access. In 1938, as-Suwaidi aspired to shift the bilateral boundary south, to incorporate the islands and adjacent sea lanes, in order to accommodate a new Iraqi port at Umm Qasr.[60] This effort failed so, throughout the 1940s and 1950s, Iraq repeatedly asked Kuwait to cede the islands in exchange for diplomatic recognition, boundary demarcation, or fresh water from the Shatt al-ʿArab. To further sweeten the deal, Baghdad even offered to let Kuwait maintain its oil rights in any ceded territories: further evidence of Iraq's limited petroleum ambitions.[61] Nonetheless, the Kuwaitis persistently refused Iraq's proposals, and joint boundary commissions, active from 1966 to 1967, failed to resolve the issue.[62]

The dispute intensified in the late 1960s, largely as a result of Iraq and Iran's escalating Shatt al-ʿArab disagreement, which threatened the former's Persian Gulf access. In 1969, when Iran abrogated the Tehran Treaty, claiming a thalweg boundary in the waterway and providing naval escorts for Iranian commercial ships, Iraq stationed troops in Kuwaiti territory, near Umm Qasr. After the immediate crisis passed, the state failed to withdraw its soldiers. In late 1972, after Kuwait refused Baghdad's request for a loan, Iraq stationed more troops in the border areas. The next year, Iraqi forces occupied a Kuwaiti border post at as-Samita, killing two Kuwaiti guards. Following Arab mediation, the Iraqis withdrew from that post. Yet they remained in the other border zones. Meanwhile, further negotiation efforts from 1970 to 1978 also failed to produce a territorial accord.[63]

By the end of the decade, both countries' positions on the dispute had hardened. The Algiers Agreement, which granted Iran equal sovereignty over the Shatt al-ʿArab, strengthened Iraq's determination to acquire alternative Persian Gulf access routes. However, Kuwait's increasingly antagonized National Assembly issued a resolution pledging to retain all of the state's sovereign territory. In the early 1980s, Kuwaiti authorities refused multiple Iraqi requests for an island lease so the state could enlarge its port facilities at Umm Qasr. The Kuwaitis also consolidated their symbolic authority over the contested territories by building an unnecessary bridge from the mainland to unpopulated Bubiyan in 1982 and garrisoning forces on the island in 1984.[64] Iraq was unable to react to these perceived provocations because of its dependence on Kuwait's support in the Iran–Iraq War.[65]

The conflict confirmed Iraqi fears about overreliance on the Shatt al-ʿArab. During the war, the waterway was blocked by sunken boats, and after it ended,

Iran's refusal to implement the ceasefire precluded clearing the channel.[66] The war and its aftermath also exposed insecurities in Iraq's oil transportation outlets. In April 1982, Syria closed its main pipeline, leaving Iraq with only a one-million-barrel-per-day export route through Turkey.[67] In January 1990, when Turkey interrupted the flow of the Euphrates River for a month to fill the reservoir behind the Ataturk Dam, Baghdad could not protest, because it did not want to threaten this pipeline route.[68] These constraints increased Iraq's commitment to enlarging and diversifying its oil export facilities on the Gulf coast and securing adjacent sea lanes. Yet border discussions with Kuwait in August 1988 and February 1989 again failed to produce an accord.[69]

By 1990, Kuwait and Iraq had built up intense bilateral acrimony over the boundary issue, which increased the appeal of Iraqi aggression. It appeared more justified, given Iraq's historical claims to the neighboring territory and Kuwaitis' apparent indifference to Iraq's economic security. It also meant that a successful occupation would generate additional, nonoil benefits. It would convey island authority, which would improve Iraq's Persian Gulf access by creating a viable alternative to the Shatt al-ʿArab.[70] It would also enhance the regime's domestic standing, as Saddam could portray himself as the "liberator of usurped Iraqi lands."[71] Yet, despite these added incentives, coupled with relatively low invasion, occupation, and international obstacles from local countries, Saddam still initially refrained from attacking Kuwait.

The Iraqi leader's hesitation arose from his fear of international retaliation by the United States. Contrary to the conventional interpretation of Iraq's invasion, which asserts that Saddam attacked after receiving a "green light" from Ambassador Glaspie, the Iraqi leader was aware that the US government would retaliate, economically and militarily, for his invasion.[72] There was significant precedent for economic punishment; Iraq was already suffering under some US trade restrictions, which would likely be extended to Iraqi and Kuwaiti oil if he launched an attack. Iraqi leaders also entertained no doubts about a US military response to their aggression. In January 1991, on the eve of the coalition air campaign, ʿAziz told Baker, "We have been expecting US military action against Iraq. . . . This conduct on our part wasn't the result of ignorance."[73] Later, ʿAziz rejected the "green light" interpretation of the Glaspie meeting, stating, "She didn't tell us . . . that the Americans would not retaliate. That was nonsense you see. It was nonsense to think that the Americans would not attack us."[74]

Given the likelihood of US retaliation, the outcome of an invasion was uncertain at best. It is therefore unsurprising that Iraq initially refrained from attacking its neighbor, instead responding to its escalating crisis with alternative, peaceful activities. As noted earlier, one of these was domestic economic reforms, which failed to improve the regime's economic standing. Another was international diplomacy. This strategy had two targets: Iraq's Arab creditors and the United States. From late 1989 through mid-1990, Iraqi officials repeatedly approached Saudi Arabia, Kuwait, the UAE, and Qatar, trying to persuade them to cancel Iraq's debts, offer the state new loans, and abide by their OPEC production quotas. Iraqi authorities rationalized the first request, to cancel their war debts, by arguing that Iraq had been fighting its war with Iran on behalf of all Arabs, in order to resist Tehran's threatening, revolutionary regime. Moreover, the Iraqis argued, the region's other oil-producing states had already reaped enormous economic benefits from the war. Since Iraqi and Iranian oil output had dropped during the conflict, other producers had dramatically increased their resource revenue by making up the difference. Iraq had therefore already repaid its debts.[75]

Some of the Gulf states responded to these diplomatic entreaties with concessions. Saudi Arabia canceled most of Iraq's war debts, partly in exchange for a nonaggression pact. Kuwait, however, was more recalcitrant, despite repeated Iraqi requests. These were initiated at an OPEC meeting in November 1989, where Iraq asked that the price of oil be raised to $21 per barrel, with the promise that it not fall below $18. Since Kuwait's oil minister did not respond to this appeal, Saddam sent another request directly to the country's ruler, Emir Jabir. The emir assured the Iraqi president that Kuwait would abide by its OPEC quota but soon violated his pledge by continuing to overproduce. In January 1990, when Iraq's deputy prime minister, Saʿdun Hammadi, visited Kuwait to request debt forgiveness and a $10 billion reconstruction and development loan, the emir offered only a $500 million loan and did not reduce national oil output.[76]

After an Arab Cooperation Council (ACC) summit in February, where Iraqis asked various Gulf state leaders to forgive their state's war debts and provide another $30 billion in loans, Iraq's oil minister, Isam al-Chalabi, personally delivered the same message to Jabir. He also asked the Kuwaiti leader to abide by his state's production quota, with no tangible results. At an OPEC meeting in May, al-Chalabi again pushed for members to respect their quotas in order to keep oil prices above $18 per barrel. Kuwait announced some cuts but continued to overproduce.[77] Later that month, the tone of Iraqi appeals became more heated. At a special Arab summit in Baghdad from May 28 to May 30, Saddam aired his grievances against Kuwait and the UAE. In a private session, he accused the states of deliberately undermining Iraq through their overproduction. For every $1 drop in the price of oil, he claimed, Iraq lost $1 billion in annual revenue. Saddam equated overproduction's impacts with warfare: "We say that war is fought with soldiers and much harm is done by explosions, killing, and coup attempts—but it is also done by economic means. . . . Therefore, we would ask our brothers who do not mean to wage war. . . . This is in fact a kind of war against Iraq. Were it possible, we would have endured. But I believe that all our brothers are fully aware of our situation. . . . We have reached a point where we can no longer withstand pressure."[78]

Despite the limited effectiveness of these diplomatic endeavors, Iraqi officials persisted in them for another eight weeks. In late June, an Iraqi delegation, including Hammadi, traveled to various Gulf states to discuss oil production and loans. At their meetings in Kuwait, the Iraqis asked for a $10 billion loan and quota adherence. The Kuwaitis again refused to forgive Iraq's debts and offered only another $500 million, dispersed over three years.[79] On the oil production issue, Kuwait refused to lower its output and may have implied that it had the right to increase it. Historians Majid Khadduri and Edmund Ghareeb write that "Hammadi, rightly or wrongly, seems to have gotten the impression . . . that Kuwait would not be bound by the OPEC quota."[80] On July 10, oil ministers from Saudi Arabia, Iraq, Kuwait, and the UAE finally agreed to an $18 price target. However, the Kuwaitis again proved unreliable. A day after promising to rein in production, Kuwait's oil minister said that his state would review and potentially revise its position in the fall. He also suggested that Kuwait would propose eliminating the quota system entirely at OPEC's meeting in October 1991.[81]

Given this resistance, "by mid-July Baghdad felt it had almost exhausted diplomatic methods of resolving the dispute."[82] On July 16, ʿAziz submitted a memorandum to the chairman of the Arab League (dated July 15), detailing Iraq's complaints against its neighbor. The document claimed that "the officials of the government of Kuwait . . . have attempted in a planned, predetermined, and continuous process to take advantage of Iraq and to cause it harm with the intention of weakening it after the end of the ruinous war which lasted eight years. . . . This policy was pursued out of selfish and narrow interests and goals which we cannot any longer but consider as suspicious and dangerous." The memorandum characterized Kuwait's overproduction as "a planned operation" which, for Iraq, "means a loss . . . of several billion dollars in revenue this year at a time when it is suffering from a financial crisis because of the costs of its rightful defense of its own land." The memorandum reiterated that Kuwait's overproduction and theft were "tantamount to military aggression" and mentioned that, when Iraq had raised the issue before, Kuwait and the UAE had responded with "insolent statements." For the first time, Iraq brought the Rumaila oil field into the dispute, accusing the Kuwaitis of slant drilling into the reservoir and siphoning off Iraqi reserves. The memorandum asked Arab League members to persuade Kuwait and the UAE to change their behaviors and implement a price increase to $25 per barrel. It also again requested debt cancellation and demanded compensation from Kuwait for the $2.4 billion of oil that it had purportedly "stolen" from Rumaila during the Iran–Iraq War.[83]

Kuwait's July 19 response to Iraq's memorandum was intransigent. The state denied sole responsibility for the oil price collapse, defended its right to drill in Rumaila, and offered no financial assistance or promises on oil output.[84] Iraqi leaders interpreted this obduracy as a sign of the United States' involvement in the price collapse. Their reasoning was deductive; they assumed that, without American support, such a small, militarily weak state would not willfully defy all of Iraq's demands.[85] As the July 15 memorandum had asserted, "We can only conclude that those who adopted this policy directly and openly and those who supported it and pushed for it were carrying out part of an imperialist Zionist plan against Iraq and the Arab nation."[86] In their July 21 rejoinder to Kuwait's response, the Iraqis were even more explicit in their accusations, asserting that "the Kuwaiti Government's policy was a US policy."[87] The announcement, a week later, of joint naval exercises between the United States and the UAE appeared to confirm Iraqi suspicions, as did the movement of US ships and KC-135 aerial tankers to positions closer to Kuwait and the UAE.[88]

Believing that the United States was the driving force behind Iraq's intensifying crisis, Saddam's diplomatic initiatives also targeted US officials. ʿAziz had initially aired the regime's grievances to Baker in October 1989. In February 1990, Saddam lectured Assistant Secretary of State John Kelly about the post–Cold War geopolitical situation in the Persian Gulf, attempting to persuade him that the United States should use its newfound "free hand" for "constructive purposes" rather than "blindly" following Israel.[89] When Iraq hosted a congressional delegation led by Senator Robert Dole in April, Saddam again highlighted his fear of a US-backed Israeli attack and attempted to clarify that his "burn half of Israel" speech had been intended as a deterrent. He asserted that Iraq preferred good relations with the United States, but only if Americans felt the same way.[90]

On July 25, following the announcement of US–UAE naval maneuvers, Saddam summoned Glaspie for their infamous private meeting. In the meeting, the Iraqi leader spoke at length about the economic crisis facing his state and his belief that the United States was inciting Kuwait's and the UAE's behavior. Glaspie viewed these sentiments as sincere; in 1991, after Operation Desert Storm, she observed that Iraqi officials were "quite convinced the United States . . . was targeting Iraq. They complained about it all the time. . . . Day after day, the Iraqi media since February [1990]—literally every day—was full of these accusations. And I think it was genuinely believed by Saddam Hussein."[91] During the meeting, Saddam also warned the ambassador that the United States should not "force Iraq to the point of humiliation at which logic must be disregarded."[92]

Over the previous ten days, Iraq had already begun moving troops to its border with Kuwait. During his meeting with Glaspie, Saddam claimed that the mobilization was a tactic to persuade Kuwait to change its behavior; "How else can we make them understand how deeply we are suffering?" he queried.[93] The mobilization caught the attention of Arab leaders, provoking a flurry of further negotiations. These initially appeared to bear fruit; after meeting with Saddam on July 24, Egyptian president Hosni Mubarak left with the impression that the crisis would soon blow over.[94] In addition, at an OPEC meeting in Geneva from July 26 to July 27, the organization's member states, including Kuwait and the UAE, finally agreed to adhere to their production quotas in order to achieve a $21 per barrel oil price.[95] Yet Saddam had little faith that the Kuwaitis would keep their word, and a subsequent meeting in Jiddah on July 31 gave little reason for optimism.[96]

In addition, following Saddam's meeting with Glaspie, the United States did not back down. Rather than attempting to reassure Iraq of its benign intentions, on July 27, the Senate voted to block any further CCC guarantees to Iraq and all deliveries of militarily useful equipment.[97] Saddam's conviction that the United States was determined to defeat him appeared to be confirmed. The credit cutoff, coupled with Iraq's escalating budgetary crisis, meant that the state literally could no longer feed its population.[98] Meanwhile, the regime had exhausted all alternative means of improving its situation. Domestic economic reforms had merely intensified popular animosity, while international diplomacy had failed to eliminate Iraqi war debts, increase oil prices, or alleviate US hostility. Saddam therefore turned to foreign aggression; Iraq invaded Kuwait on August 2.

Before launching the invasion, Saddam was uncertain about its outcome. The Iraqi president knew that seizing Kuwait would be easy, militarily. Yet he anticipated US retaliation and recognized that American forces severely outmatched his own. Saddam hoped, nonetheless, that the United States would prove to be a paper tiger. He told Glaspie that the American public was unwilling to stomach significant casualties. In addition, his speech before the ACC in February had highlighted the United States' withdrawal from Lebanon in 1983, following bombings of the US embassy and Marine Corps barracks that killed over two hundred soldiers, as evidence of this weakness.[99] Saddam conjectured that, if Iraqi forces could kill a sufficient number of Americans, the United States might forgo further military involvement in Kuwait and Iraq.

To increase the likelihood of Iraqi military success, in late July, Saddam decided to seize the entirety of Kuwait, instead of only the contested northern border regions of Ritqa and Qasr, which contained the Rumaila oil field and the disputed islands.[100] The Iraqi leader believed that expanding the scope of the invasion would impede a US response, as he mistakenly assumed that Saudi

Arabia would not permit US troops to be stationed in its territory.[101] In addition, to eliminate one incentive for American retaliation, after the invasion, Saddam repeatedly offered to sell the United States and other consumer states Kuwaiti and Iraqi oil for less than $25 per barrel.[102] He hoped that, if Iraq demonstrated that it was still a reliable petroleum supplier, third-party states might accept the occupation.

Even if this strategy failed and the United States proved to be a more committed adversary, the possibility of losing to the Americans was a risk that Saddam and other Iraqi leaders were willing to take. If Iraq failed to act, they believed, the Baʿathist regime would inevitably collapse as a result of domestic instability or international attacks perpetrated by Israel and the United States. The Iraqis therefore perceived their invasion as defensive; as Saddam stated later, it was a means of "defending by attacking."[103] In meetings after the conflict began, the president also insisted that he had no other choice if he wanted to save his state.[104] Speaking to Yemeni president ʿAli Abdullah Saleh on August 4, Saddam observed that his regime could not survive without supplying food and other public goods to its people. He also asserted that Iraq had exhausted all peaceful, diplomatic methods of persuading Kuwait to alter its behavior. In a meeting with Russian special envoy Yevgeny Primakov on October 6, Saddam claimed that Iraqis had been backed into a corner, leaving an invasion "the only choice we had."[105] He and his advisers also blamed the United States for precipitating the confrontation. In ʿAziz's words, "We were pushed into a fatal struggle in the sense of a struggle in which your fate will be decided. You will either be hit inside your house and destroyed, economically and militarily. Or you go outside and attack the enemy in one of his bases. We had to do that, we had no choice, we had no other choice. Iraq was designated by George Bush for destruction, with or without Kuwait."[106]

Iraqi authorities hoped that, by occupying Kuwait, they could "change the balance of power in [their] favor."[107] However, this aspiration was soon dashed. The UN quickly imposed sanctions on sales of Iraqi and Kuwaiti oil, and the Saudis allowed Western troops to operate in their territory. After an exhaustive targeted bombing campaign in January 1991, coalition ground forces began to engage Iraqi troops in mid-February. By the end of the month, they had pushed Iraqi forces out of Kuwait and declared a ceasefire. However, President George H. W. Bush decided not to continue the counterattack, allowing Saddam to remain in power. Noting the United States' failure to occupy Iraq or overthrow him, the Iraqi leader concluded that he had won the war.[108] This conviction is counterintuitive to outside observers, partly because it calls classic oil war interpretations of the invasion into question. If oil was a significant target of Iraqi aggression, the attack failed miserably. Iraq did not increase its petroleum resources or revenue and, from an oil standpoint, was worse off after the war than it had been before it, because of the international economic sanctions that curtailed Iraqi petroleum sales until the mid-1990s. However, if we recognize that Saddam's fundamental goal was survival, not oil, his assertion becomes more credible.

Iraq's original invasion plan, which targeted only Kuwait's northern provinces, also undermines classic oil war arguments. If petroleum was a central goal of the Iraqi attack, Saddam should have had more expansive invasion plans from the beginning. Instead, Iraq's war planning left much of Kuwait's petroleum untouched. This restraint suggests that the state had limited interest in its neighbor's oil per se. Instead, the invasion had a different aim: sustaining the regime's survival. Saddam believed that attacking Kuwait and controlling some of its resources might help him achieve that end. Hence, the oil grab was a gambit, aimed at accomplishing a different, broader goal.

Should we nonetheless label oil gambits classic oil wars, since these attacks target petroleum resources? If we do, we must also acknowledge that these conflicts look very different from what many of us have assumed. Oil gambits are not launched by greedy states attempting to increase their wealth by acquiring foreign petroleum resources. Instead, they are perpetrated by desperate states who view these resource grabs as their only possible means of survival.[109] Leaders initiate these conflicts reluctantly, after exhausting all other means of satisfying national oil needs. And they are still selective in their targeting, reserving international aggression for dispute scenarios. Finally, oil gambits are exceedingly rare; only one has occurred since petroleum became a critical strategic resource. Altogether, the Iraqi case—and the lack of additional, historical oil gambits—provides further evidence of countries' disinclination to fight for oil.

Postscript: The US Invasion of Iraq (2003)

In March 2003, the George W. Bush administration accomplished what the first Bush administration had not: US forces advanced to Baghdad and overthrew Saddam Hussein. Many observers assumed that Operation Iraqi Freedom was a classic oil war. A Pew Research Center survey conducted in late 2002 revealed that 75 percent of French respondents, 76 percent of Russian respondents, and 54 percent of German respondents believed that the United States was planning to invade Iraq because it "wants to control Iraqi oil."[110] This belief was even more prevalent in the Arab world; 83 percent of people surveyed in Jordan, a US ally, agreed with the statement.[111] Members of the Bush administration, however, rejected oil war interpretations of the conflict. Secretary of Defense Donald Rumsfeld notoriously asserted that the invasion "has nothing to do with oil, literally nothing to do with oil."[112] David Frum, one of Bush's speech writers, also claimed,

"The United States is not fighting for oil in Iraq."[113] On the other side of the Atlantic, Prime Minister Tony Blair insisted, "The war in Iraq has nothing to do with oil, not for us, not for the UK, not for the United States."[114]

These official claims are questionable, given long-standing British and American interests in the Persian Gulf's petroleum resources. Nevertheless, I agree that the 2003 invasion was not a classic oil war. The Bush administration did not overthrow Saddam to obtain control over Iraq's petroleum resources. Nor did the United States invade Iraq to benefit American and British oil companies.[115] This does not mean, however, that the invasion was entirely divorced from petroleum concerns. In particular, US officials' desire to increase global oil output, coupled with their fear that a revenue windfall would empower Saddam, may have contributed to their pursuit of regime change. However, as Greg Muttitt observes, fighting to facilitate the expansion of another country's oil output is different from "want[ing] *the oil itself* as some form of imperial plunder" (emphasis in original).[116]

There is abundant evidence that the United States was not prosecuting a classic oil war in Iraq. First, the Bush administration's prewar planning contradicted this aim. To exploit Iraq's oil resources, it would be necessary to occupy large sections of the country over the long term. However, the Bush administration expected to begin drawing down American forces within a few months of the invasion and depart entirely in less than a year.[117] US planning documents also envisioned a speedy restoration of local control over the Iraqi oil industry. In September 2002, Douglas Feith, the undersecretary of defense for policy, created the Energy Infrastructure Planning Group (EIPG) to "develop a comprehensive contingency plan for protecting, repairing and operating Iraqi energy infrastructure." The planning guidelines produced by the EIPG asserted that "Iraqi petroleum resources belong to the people of Iraq," that oil production "should involve existing Iraqi personnel and organizations," and that the United States would restore "production and marketing responsibilities to a stable Iraqi authority as soon as practicable."[118] Since these classified briefings were never intended for public consumption, it is likely that they reflected the administration's actual intentions. When President Bush was briefed on the petroleum plans in February 2003, he agreed that the United States should "give them [Iraqis] full control as soon as possible."[119]

At first glance, the United States' behavior during the invasion appears to belie these commitments. Military planners prioritized the seizure of oil installations; Iraq's Persian Gulf export terminals and the Rumaila oil field were some of their preliminary targets.[120] In addition, after entering Baghdad, US forces notoriously defended the Ministry of Oil, while leaving other government buildings unguarded.[121] However, these initiatives do not prove that the Bush administration was attempting to acquire long-term control over Iraq's oil resources. Given the state's overwhelming dependence on petroleum revenue, failing to immediately secure its oil infrastructure would have provoked accusations that US forces had endangered Iraq's national patrimony, especially since rapid production shutdowns could permanently damage oil reservoirs.[122] Moreover, within a few months of the invasion, it became evident that, if the United States' top priority was controlling Iraq's oil, it had bungled the job. After their initial advance, US troops failed to defend many of the facilities they had captured, allowing extensive looting.[123] Three months after the invasion, key sites remained unsecured.[124]

Second, the United States' administrative choices after toppling Saddam fail to support classic oil war arguments. Although Iraqi exiles participated in the prewar planning process, after the invasion, General Jay Garner, the leader of the original US occupying authority, selected Thamir al-Ghadhban, an internal Iraqi technocrat, as interim oil minister. Al-Ghadhban had served as the Ministry of Oil's planning director before the war and was unlikely to defer to US preferences regarding the management of Iraqi petroleum resources. The Bush administration also appointed Philip J. Carroll, the former CEO of Shell USA, as senior American adviser to the oil ministry. Carroll resisted efforts to de-Baʿathify the ministry, insisted that Iraqis retain ultimate decision-making authority over their oil, and ignored calls to privatize the industry. "I told everyone that I would have no part of it," Carroll stated. "For 25 million people to lose control of the one thing they have that is of value would be highly irresponsible."[125] The Bush administration accepted this approach; in September 2003, when the Coalition Provisional Authority legalized foreign ownership of most of Iraq's public companies, it omitted the oil industry.[126] US government and oil company representatives subsequently encouraged Iraqi officials to accept greater foreign investment in their petroleum projects, in order to increase resource output.[127] However, the state retained ultimate control over its oil and investment decisions.

The Bush administration's willingness to rapidly restore local authority over the national government and oil industry may have reflected its awareness of the obstacles to classic oil wars. In 1991, members of the previous Bush administration had refrained from invading Iraq and overthrowing Saddam because they recognized that there would be intense international opposition to expanding Operation Desert Storm beyond its original mandate of defending Kuwait. Officials also anticipated that regime change would lead to a prolonged US occupation, provoking intense local resistance.[128] As Paul Wolfowitz, then undersecretary of defense for policy, stated, "A new [regime] in Iraq would have become the United States' responsibility. Conceivably, this could have led the United States into a more or less permanent occupation of a country . . . where the rule of a foreign occupier would be increasingly resented."[129]

Many of the US officials involved in the 1991 decision, including Wolfowitz, Dick Cheney, and Colin Powell, became prominent members of the George W. Bush administration. They were therefore cognizant of the obstacles to conquering Iraq and seizing its petroleum resources.[130] The EIPG's briefings, which were presented to National Security Council deputies, also highlighted these impediments.[131] The group emphasized the risk of invasion obstacles, noting that, if Iraqis defended their oil installations, "battle damage and collateral damage could be significant." The EIPG also anticipated deliberate sabotage of oil facilities and cautioned that US engineers should be prepared to "fight over 1000 well fires, if necessary."[132] President Bush expressed similar concerns, observing, as the invasion began, that "if they really blow them [the oil wells], it will be years" before they could operate.[133] Accordingly, the administration assumed that conquering Iraq would result in a short-term drop in the country's oil output. These losses would be compounded if Saddam launched missiles at Kuwaiti or Saudi oil facilities in retaliation for a US attack, as some officials expected.[134]

The administration also recognized the international and occupation obstacles to grabbing Iraq's oil. Rumsfeld later recalled that he "was concerned that people across the Muslim world would believe that the United States sought to establish a colonial-type occupation for the purpose of taking Iraq's oil."[135] The EIPG noted that some American activities, such as aggressively increasing Iraq's oil output or using resource revenue to pay for the occupation, would be "highly controversial," as they would substantiate popular suspicions "that [the] incursion is driven by oil considerations." The group warned that these activities "could generate domestic Iraqi opposition."[136] A study cosponsored by the Council on Foreign Relations and the James A. Baker III Institute for Public Policy reiterated this concern, stating, "If the United States appears to be taking over Iraq's oil sector, guerilla attacks against U.S. military personnel guarding oil installations are likely."[137]

The EIPG and Council on Foreign Relations/Baker report encouraged the Bush administration to engage in aggressive public diplomacy to counter these perceptions in order to moderate local and international opposition to the invasion. As the latter stated, the United States must "reassure Iraqis and the international community about the limited nature of its intentions" to offset "the widely held view that the campaign against Iraq is driven by an American wish to 'steal' or at least control Iraqi oil."[138] The EIPG proposed a number of themes for public diplomacy, including "We want to work with the Iraqis themselves and the international community in administering petroleum proceeds for the benefit of the Iraqi people" and "We will not administer oil assets on a long-term basis."[139] However, administration principals recognized that the best way to limit local and international retaliation was to quickly restore Iraq's authority over the industry. They proceeded accordingly.[140]

Lastly, the Bush administration was keenly aware of the investment obstacles to seizing Iraq's oil. Vice President Dick Cheney, in particular, had extensive knowledge of the global oil industry, based on his experience as CEO of Halliburton, a major oil services company. He and other US officials knew that international oil companies would not invest in Iraqi oil projects until the country had a stable government that could issue legitimate contracts. Any agreements established with an interim, US-led administration could later be challenged, potentially resulting in major financial losses. Representatives of BP explicitly told the British government that they "would not wish to be involved in opaque, ambiguous arrangements" that preceded the creation of a sovereign Iraqi government.[141] Phil Carroll concurred; foreign companies "will want to see an Iraqi government, and have confidence in it, before sinking down large sums of money."[142] The only way to attract investment capital to Iraq's oil industry was to let Iraqis control it.

Given these invasion, occupation, international, and investment obstacles, prosecuting a classic oil war in Iraq would have been a "gamble of enormous proportions," and there is no compelling evidence that the United States was attempting one.[143] It is also implausible that the invasion aimed to benefit American and British oil companies.[144] The United States did award a massive, no-bid contract for postwar petroleum infrastructure repairs to Kellogg, Brown and Root (KBR), a Halliburton subsidiary. However, this decision was pragmatic, not an attempt to enrich Cheney's former employer. Very few firms were capable of executing the reconstruction project, KBR was already a preferred contractor for the US Army's Logistics Civil Augmentation Program, the company's employees possessed the security clearances necessary to work on a classified project, and the administration did not have time to complete a competitive bidding process before the invasion.[145] Over the long term, however, the United States would not be able to compel a sovereign Iraqi government to preferentially issue contracts to American and British oil companies. Instead, Iraqi officials would select the corporate partners that offered them the best financial terms and technological prowess—if they allowed foreign oil companies to participate in their industry at all.

The Bush administration did not invade Iraq to control its oil.[146] Nor was it trying to grab Iraq's petroleum resources for US oil companies. Yet that does not mean that the 2003 invasion was divorced from petroleum objectives. Oil has been one of the US government's core interests—if not its primary interest—in the Persian Gulf since the 1940s.[147] Were it not for the region's petroleum resources, it is unlikely that the United States would have intervened in Iraq in 1991 or 2003.[148] In addition, in 2003, the Bush administration had a specific, oil-related incentive to overthrow Saddam.[149] In the early 2000s, the United States was confronting an apparent energy crisis. California began to experience rolling blackouts in

June 2000 and, the following year, several prominent policy groups reported that, at current production rates, global oil output would not keep pace with rising demand.[150]

Iraq offered an escape from this resource dilemma. If the international community lifted economic sanctions, the state could accept the foreign capital and materials it required to significantly boost its oil output and help satisfy global petroleum demand.[151] Yet the Bush administration was loath to pursue this strategy, as officials believed that Saddam would use the revenue windfall from increased oil sales to accelerate his development of WMD and renew his attacks on his oil-endowed neighbors. As Cheney vividly opined in August 2002, "Armed with an arsenal of these weapons of terror, and seated atop 10 percent of the world's oil reserves, Saddam Hussein could then be expected to seek domination of the entire Middle East, take control of a great portion of the world's energy supplies, directly threaten America's friends throughout the region, and subject the United States or any other nation to nuclear blackmail."[152]

Regime change would remove this obstacle. A less hostile Iraqi government could collect more oil revenue without threatening regional or US security. Deposing Saddam would therefore help the United States advance its core oil interest in the Persian Gulf: maintaining a steady flow of affordable petroleum supplies. Once that was accomplished, US forces could withdraw from the country. There was no need to retain direct, sustained control over Iraq's oil resources or guarantee their extraction by US oil companies; the Bush administration merely needed to install a competent local authority. Secretary of State Colin Powell summarized this strategy when he asserted, in July 2003, "We have not taken one drop of oil for U.S. purposes. Quite the contrary. We put in place a management system to make sure that Iraqi oil is brought out of the ground and put onto the market."[153]

Debate persists about whether the United States actually invaded Iraq to increase the country's oil output. Some authors claim that this was the Bush administration's primary goal, while others, such as F. Gregory Gause III, assert that "there is no evidence from the public record that oil considerations played [this] role . . . in the Bush administration's decision to go to war."[154] Even if the United States did launch the invasion for this reason, the case does not contradict the book's central finding: that states are reluctant to fight for petroleum resources. Even unrivaled superpowers avoid classic oil wars.

Conclusion

PETRO-MYTHS AND PETRO-REALITIES

Despite oil's extraordinary economic and military value, countries have largely refrained from international petroleum grabs. They have avoided classic oil wars. Only one state has launched an oil gambit. A few countries have initiated oil campaigns, in the midst of ongoing international wars that were started for other reasons. States have perpetrated fewer than twenty oil spats. These numbers are remarkably small, particularly in comparison to the number of countries that could have fought to obtain petroleum resources over the course of almost a century. Moreover, countries initiated their oil spats selectively, reserving them for situations in which obstacles were limited or additional gains were large. In the severe conflicts—the oil campaigns and oil gambit—leaders exhausted all other means of satisfying national petroleum needs before turning to international aggression. And in all the conflicts, petroleum ambitions were never decision makers' sole motive for aggression. States are evidently extremely reluctant to fight for oil resources.

My findings controvert the oil wars myth, as well as popular interpretations of many of the twentieth century's deadliest international conflicts. They also challenge international relations scholars' assumptions that fighting for oil pays and that the resource is a significant cause of interstate conflict. In addition, these results raise a number of further questions. First, will countries fight to control petroleum resources in the future if oil prices rise or fall? Second, why is oil different from other natural resources that were the objects of imperialist projects? Third, how does the divergence between the oil wars myth and the historical record matter? Fourth, does states' reluctance to prosecute classic oil wars, oil gambits, oil campaigns, and oil spats affect their willingness to engage in other forms of oil-related contention? And fifth, what does this mean for US foreign policy? This conclusion will begin to answer these questions, while encouraging other researchers to pursue them further.

Same as It Never Was

As oil prices climbed in the early 2000s, commentators issued increasingly dire predictions of incipient "peak oil": the apex of global petroleum production, followed by an inevitable decline in resource output and concomitant jump in oil prices.[1] However, after 2014, when oil prices plummeted, the discourse quickly flipped to predictions of "peak oil demand": the apex of global petroleum *consumption*—precipitated by improving fuel efficiency, electrified transportation, and a transition to renewable energy sources—which would cause oil prices to stagnate or decline.[2] Thus, in less than twenty years, we have anticipated two dramatically different energy futures. Yet neither of these trajectories is likely to alter the frequency of conflicts over oil resources. Countries refrained from classic oil wars when prices were below $10 per barrel, and they avoided them when prices soared above $145. There is no reason to expect their behavior to change in the future. States will continue to eschew classic oil wars and red herrings will remain the dominant form of conflict in hydrocarbon-endowed territories.

Countries will also continue to prosecute oil campaigns in the midst of ongoing wars as long as their militaries run on petroleum-based fuels. However, these larger conflicts will not be driven by countries' oil ambitions; instead, if history is an accurate guide, they will be provoked by hegemonic aspirations. China is therefore the most likely future oil campaigner. If China and the United States become involved in a hegemonic war, the latter may try to interrupt the former's oil access in order to obtain a military advantage.[3] China may respond by launching an oil campaign, most likely targeting central Asia or Siberia, after exhausting all other means of meeting its wartime petroleum needs.

Oil spats will also remain a persistent feature of international politics, regardless of changes in resource prices. Rival states, such as China and Japan, will be particularly prone to these conflicts, as they obtain additional benefits from petroleum sparring. Rising oil prices could increase the frequency of oil spats, as these conflicts will appear marginally more beneficial, while entailing the same, relatively low costs. However, declining oil prices could also precipitate more oil spats, as falling oil revenue will incite popular discontent in many petroleum-producing states, encouraging governments to engage in diversionary activities, including oil sparring. Venezuela's president, Nicolas Maduro, has already attempted this maneuver, decrying ExxonMobil's development of recently discovered Guyanese oil fields, partly to distract Venezuelans from their country's economic meltdown.[4] The frequency of future oil spats is therefore uncertain. However, regardless of their number, these mild incidents will not threaten international security, as they consistently fail to escalate.

Oil gambits are far more dangerous but will remain extremely uncommon. The one historical case, Iraq's invasion of Kuwait, demonstrated that an exceptional constellation of circumstances is required for a state to initiate an oil gambit. First, a prospective aggressor must believe that it faces an existential threat that it could resist by seizing foreign petroleum resources. Second, it must possess a viable target: a neighboring, oil-producing state that it can defeat militarily. Third, it must have exhausted all other means of increasing its petroleum resources or revenue before initiating its attack. These criteria are not particularly sensitive to oil prices, so oil gambits will remain rare and desperate events.

These patterns will only change if the oil wars myth begins to drive leaders' decision making. If officials believe that countries fight wars to obtain petroleum resources, the narrative could become a self-fulfilling prophecy.[5] Anticipating that other countries will perpetrate international oil grabs, states may engage in increasingly mercantilist activities to secure their access to petroleum supplies. In doing so, they will restrict global trade and antagonize other countries, heightening international tensions and limiting states' ability to peacefully satisfy their resource needs. Under these circumstances, governments may eventually be compelled to initiate classic oil wars. In contrast, questioning the oil wars myth will help states resist these autarkic urges, reducing the risk that they will later have to fight for oil resources.

Is Oil the Exception?

As I noted in chapter 1, many authors claim that oil is exceptional, in the sense that it is the one natural resource that countries will fight for.[6] However, my analysis revealed that states may be *less* willing to fight for petroleum than for other resources. Oil has not inspired the same imperial adventurism that gold, spices, salt, and iron provoked in the sixteenth to nineteenth centuries. The reasons for this discontinuity, I argue, are the characteristics of the oil industry and timing. By 1912, when petroleum became valuable enough to potentially fight over, states already faced substantial obstacles to seizing foreign oil resources. Nationalism had become a potent force in international politics, so local populations were likely to resist foreign rule, heightening occupation obstacles. In addition, emerging norms against conquest, plunder, and the forceful resolution of interstate disputes created international obstacles to petroleum grabs, as third parties were increasingly likely to retaliate for these actions.

The physical and political economic characteristics of the oil industry also discouraged imperial adventurism. Petroleum exploration, production, and transportation have always required extensive, expensive physical infrastructure, which can be damaged by military aggression. The oil industry has also persistently relied on access to large amounts of foreign capital, which investors are likely to withhold from conquered territories. Hence, the invasion, occupation, international, and investment obstacles to fighting for petroleum have always been high. Unlike other resource wars, classic oil wars have never paid.

Other characteristics of the modern international system have also discouraged oil grabs. By the beginning of the twentieth century, states had greater technological capacity to develop petroleum substitutes—although, as chapter 7 illustrated, some of them have been more successful at this than others. More importantly, the international economic system has allowed states to buy oil rather than fighting for it. This trading system has been imperfect. The world wars restricted countries' access to oil resources and revenue. So did peacetime trade restrictions, such as the oil embargo imposed by Arab members of OPEC in the 1970s and international sanctions against Iraq in the 1990s. Nonetheless, the market has generally been remarkably effective at satisfying countries' petroleum needs.[7] Moreover, as the number of oil-producing states has risen, the viability of supply shutoffs—and the legitimacy of oil grabs—has declined even further.[8] It is increasingly difficult for a government to claim that it had "no other choice" but to seize another country's petroleum resources. Thus, while oil is exceptionally valuable, it is currently unnecessary for states to treat it differently from other commodities. It is quantitatively, not qualitatively, distinct from other natural resources.

Meanwhile, research on other resources suggests that states' reluctance to fight for oil may not be exceptional. Aaron Wolf has found that international "water wars"—another type of resource conflict that figures prominently in popular and academic narratives—are far rarer than most people assume. Between 1918 and 1994, states fought no wars for water and only seven water skirmishes: the conceptual equivalent of oil spats.[9] This raises another question: If states refrain from wars to acquire the "water of life" and from wars to obtain "the lifeblood of industrially advanced nations," which natural resources *do* they fight for? Perhaps we have overestimated all resources' contributions to interstate conflict. Scholars should examine this question more closely rather than simply assuming that countries' desire to obtain valuable resources has been—and continues to be—a significant cause of international conflict. In doing so, they should employ methods that go beyond correlation, so they can determine whether countries actually fought *for* natural resources or merely fought *in* resource-endowed territories.

Moving forward, we should also adopt a more skeptical view toward future resource war claims. Inevitably, countries will confront new resource shortages, and given the Mad Max and El Dorado myths' durability, we should expect these narratives to remain culturally accessible whenever the next crisis emerges. However, we should refrain from assuming that value will automatically lead to violence. States may, in fact, be equally reluctant to fight for other resources.

Believing Dangerously

In the introduction to his resource biography, *The Age of Oil*, Leonardo Maugeri observes that "throughout its history, 'black gold' has given rise to myths and obsessions, fears and misperceptions of reality, and ill-advised policies that have weighed heavily on the world's collective psyche."[10] In addition to the oil wars myth, petroleum has provoked six myths related to the 1973–1974 oil embargo, nine myths associated with the United States' unconventional "oil boom," "mythmaking on the Saudi frontier," the "myth of the Caspian 'great game,'" "myths that make Americans worry about oil," the "myth of petroleum independence," and the "myth of the oil crisis."[11] Oil apparently encourages mythmaking.

However, while many scholars have identified and challenged oil-related myths, few have attempted to explain why oil is so readily mythologized or why petro-myths are so widely believed. Perhaps people's limited understanding of the physical properties and political economy of oil encourages uncritical acceptance of all kinds of erroneous, petroleum-related claims? Alternatively, does oil's exceptional military and economic utility encourage us to misinterpret its effects? Or have the oil industry's many disreputable activities, dating back to the practices that Ida Tarbell chronicled in her 1904 exposé of Standard Oil, led us to assume that petroleum provokes extravagant and egregious behaviors?[12] Investigating why petro-myths propagate so readily would be a valuable topic for further research.

A more pressing enterprise, however, is to consider the effects of petro-myths generally, and the oil wars myth specifically. As I observed in chapter 2, hegemonic myths shape our thinking, including our interpretations of real-world events and our policy responses. They accomplish this partly through omission. Dominant narratives are, by their nature, exclusive: "a rhetorical razor that defines included and excluded, relevant and irrelevant, empowered and disempowered."[13] Hence, the oil wars myth inevitably sidelines other stories. By attributing conflicts to countries' petroleum ambitions, we obscure the other issues, interests, and actors that may be driving them. This process may occur subconsciously, as a result of our psychological tendency to latch onto familiar narratives. Or we may use the oil wars myth to deliberately manipulate.

Actors' strategic deployment of oil war narratives was particularly evident in the Chaco case, presented in chapter 5.[14] Huey P. Long, the Bolivian and Paraguayan governments, and members of the Bolivian opposition all used this storyline to advance their parochial interests: lambasting Standard Oil, attracting international support to their side of the conflict, and undermining the Salamanca regime. The Bolivian people also eventually embraced the narrative to bring order and meaning to a catastrophic and apparently irrational event.[15] Many of these actors knew that the oil war interpretation was false. They reiterated it, nonetheless, because it served their interests.

Civil war researchers have documented a number of negative consequences arising from overreliance on resource war narratives. First, attributing conflicts to belligerents' resource ambitions causes us to overlook other motives for violence, which, in the case of intrastate conflicts, are more nuanced, and often less greedy, than resource war interpretations suggest.[16] Second, resource war narratives can misplace responsibility for violence. For example, in the civil wars literature, the narratives' emphasis on local insurgents encourages us to ignore the broader social, political, and economic structures that incentivize these actors' resource-oriented aggression.[17] Third, overemphasizing resource war narratives leads to inefficient and even counterproductive policy choices, as it leaves other, significant causes of conflict, including larger structural issues, unaddressed.[18]

Accordingly, when we hear classic oil war explanations for interstate conflicts, we should ask ourselves, What is this narrative encouraging us to ignore? Whom does it benefit?[19] And which policy choices might it provoke? Notably, an early effort to denaturalize the oil wars myth came from Michael Klare. Twenty years before publishing *Resource Wars*, he asserted that the narrative was merely an invented justification for US naval expansion. As he put it in 1981, "After several decades of uncertain purpose, the navy has finally discovered a rationale for unlimited expansion: the protection of imported raw materials."[20]

By labeling a conflict a classic oil war, authorities invoke a particular storyline: two states battling for control over petroleum resources. Everything outside of these characters and plot is erased.[21] This framing empowers national governments, in addition to national military establishments; if countries must fight for petroleum resources, the state will take the lead, possibly resorting to extraordinary measures to obtain or defend petroleum supplies.[22] Leaders may also embrace the oil war narrative to bolster their popular support. Although petroleum ambitions invite international censure, they can play well at home, because they imply that the population as a whole will benefit from international aggression, since captured oil resources and revenue can be redistributed domestically. Emphasizing an oil motive also obscures leaders' more selfish incentives for international aggression, such as inciting a rally-'round-the-flag effect.

Finally, our intellectual commitments to the Mad Max and El Dorado myths limit our understanding of historical interstate conflicts. In the Mad Max myth, "good" actors fight for their survival. In the El Dorado myth, "bad" actors fight to enrich themselves. Where, then, do we put bad actors who believe that they are fighting for their survival, like Iraq in 1990 and Germany and Japan in their World War II oil campaigns? The two hegemonic myths leave no space for that storyline. Consequently, we are faced with the dilemma of either condemning international aggression or correctly interpreting these states' motives for war. To accomplish both tasks, we must first demystify—or "demythify"—classic oil wars.

Evidence of Things Not Seen

This book has found that states avoid classic oil wars. However, this does not mean that they never engage in petroleum-related violence. As other authors have observed, there are many pathways from oil to war, most of which do not involve seizing foreign or contested oil resources.[23] Countries may fight to secure oil transportation routes: a factor that contributed to the Chaco War and both of Iraq's major conflicts, because of the country's concern about its diminutive Persian Gulf coastline. States may also fight to prevent the consolidation of control over global petroleum reserves, as occurred in Operation Desert Storm (1991). In addition, they may pursue regime change in oil-producing countries in order to alter their targets' resource behavior, as the Bush administration may have attempted to do in 2003. In addition, oil-producing countries can engage in petro-aggression: using the opportunities created by their resource revenue to finance attacks on other states.

Within countries, rebels may challenge their governments or launch secessionist conflicts to obtain petroleum resources or revenue. Oil exploitation can also finance intrastate conflicts. Additionally, oil-related violence can transcend the international–domestic divide. Petroleum-related civil wars may spill over into other countries. Foreign governments may finance insurgencies in oil-producing states or intervene directly in these contests in order to support their preferred party. Oil companies may also interfere in civil wars, if they think one belligerent will be better for business.

The good news about classic oil wars is therefore not necessarily good news about other types of oil-related conflicts. The impediments to classic oil wars—invasion, occupation, international, and investment obstacles—apply unevenly, if at all, to other kinds of contention. However, my findings do suggest that these alternative forms of petro-conflict merit further attention, to evaluate whether and how participants' oil interests actually contribute to violence. Civil war researchers, as well as this book, have already demonstrated that resource war narratives are often overstated. Other pathways from oil to conflict may be equally problematic. The prevalence of petro-myths suggests that, at a minimum, all oil–conflict narratives should be interrogated, not accepted at face value.

Over the Horizon

Do my findings warrant changes in US foreign policy? On the one hand, I have determined that competition over oil resources is not a significant threat to international security. Oil spats will not escalate. Great powers will not fight wars to obtain petroleum resources, even if oil prices rise. Oil-producing countries will not attempt to seize each other's resources. This means that the United States' original purpose for establishing a military presence in the Persian Gulf—deterring oil conquest by local or extraregional states—does not exist. On the other hand, as noted earlier, there are many other types of oil-related contention that could destabilize petroleum-producing countries. While future research may challenge these hypothesized oil–conflict connections, my analysis has revealed that states regularly fight *in* hydrocarbon-endowed territories for reasons unrelated to oil. Hence, we should expect the militarization of oil-producing areas, by international and domestic actors, to remain a regular occurrence.

The question, then, is not whether hydrocarbon-endowed territories will continue to experience violent conflict; they will, even if oil is a marginal motive for contention. The relevant policy questions are, instead, whether these conflicts will endanger global petroleum flows and whether the United States' current policy choices will moderate these disruptions. Examining the US military presence in the Persian Gulf, Eugene Gholz and Daryl G. Press have answered both questions in the negative. They argue, first, that the international oil market can compensate for most interruptions to resource flows, including those precipitated by violent conflict. Second, the United States' current forward-deployed military posture does not help it deter or respond to the three major threats to regional security: international aggression, the closure of the Strait of Hormuz, and civil unrest in petroleum-producing states. In fact, the US presence exacerbates the final problem by engendering local grievances.[24] My analysis demonstrates that the US military presence also exacerbates the first problem; in 1990, it encouraged Saddam Hussein's oil gambit by strengthening his conviction that the United States was determined to overthrow his regime. My findings therefore reinforce Gholz and Press's conclusion that the United States' interests may be better served by a more limited Persian Gulf presence or even an over-the-horizon military posture.

The broader policy recommendation that emerges from this book, however, is that decision makers should refocus their attention on other causes of interstate conflict. When we attribute violence to states' petroleum aspirations, we ignore the issues that countries actually fight over: hegemonic ambitions; perceived threats to state survival; disputed territories' other economic, strategic, and symbolic assets; political independence; and national pride. These, not oil, are the factors that fuel major international conflicts.

Notes

INTRODUCTION

1. Throughout the book, I use the terms *oil* and *petroleum* interchangeably.

2. Kissinger quoted in Ian O. Lesser, *Resources and Strategy* (New York: St. Martin's, 1989), 123; Hans J. Morgenthau, *Politics among Nations: The Struggle for Power and Peace*, 7th ed. (1948; New York: McGraw Hill, 2006), 129.

3. Bérenger quoted in Anton Mohr, *The Oil War* (London: Martin Hopkinson, 1926), 30.

4. "Undecideds," *The West Wing*, season 7, episode 8, directed by Christopher Misiano, written by Deborah Cahn, aired December 4, 2005 (Burbank, CA: Warner Home Video).

5. P. W. Singer and August Cole, *Ghost Fleet: A Novel of the Next World War* (New York: Houghton Mifflin Harcourt, 2015), 15.

6. Andrew T. Price-Smith, *Oil, Illiberalism, and War: An Analysis of Energy and US Foreign Policy* (Boston: MIT Press, 2015), 81–82; Christopher J. Fettweis, "Is Oil Worth Fighting For? Evidence from Three Cases," in *Beyond Resource Wars: Scarcity, Environmental Degradation, and International Cooperation*, ed. Shlomi Dinar (Cambridge, MA: MIT Press, 2011), 227.

7. Michael T. Klare, *Resource Wars: The New Landscape of Global Conflict* (New York: Henry Holt, 2001), 29.

8. One of this quotation's early appearances was in Frank Cleary Hanighen, *The Secret War: The War for Oil* (New York: John Day Company, 1934), 33.

9. Klare, *Resource Wars*. Other authors have followed suit, combining different scales of oil wars in their analyses. For example, Valli Koubi et al., "Do Natural Resources Matter for Interstate and Intrastate Armed Conflict?," *Journal of Peace Research* 51, no. 2 (2014): 227–43; and Philippe Le Billon, *Wars of Plunder: Conflicts, Profits and the Politics of Resources* (New York: Columbia University Press, 2012), 59–84.

10. Jeff D. Colgan, "Fueling the Fire: Pathways from Oil to War," *International Security* 38, no. 2 (2013): 147–80.

11. André Månsson, "Energy, Conflict and War: Towards a Conceptual Framework," *Energy Research & Social Science* 4 (2014): 106–16.

12. Prominent contributions include Macartan Humphreys, "Natural Resources, Conflict, and Conflict Resolution," *Journal of Conflict Resolution* 49, no. 4 (2005): 508–37; Philippe Le Billon, "The Political Ecology of War: Natural Resources and Armed Conflict," *Political Geography* 20, no. 5 (2001): 561–84; Michael Ross, "Oil, Drugs, and Diamonds: The Varying Roles of Natural Resources in Civil War," in *The Political Economy of Armed Conflict: Beyond Greed and Grievance*, ed. Karen Ballentine and Jake Sherman (New York: Lynne Rienner, 2003). For a prominent challenge to claims that oil dependence provokes intrastate war, see Benjamin Smith, "Oil Wealth and Regime Survival in the Developing World, 1960–1999," *American Journal of Political Science* 48, no. 2 (2004): 232–46.

13. On slow violence, see Rob Nixon, *Slow Violence and the Environmentalism of the Poor* (Cambridge, MA: Harvard University Press, 2011).

14. Jeff D. Colgan, *Petro-aggression: When Oil Causes War* (Cambridge: Cambridge University Press, 2013).

15. Colgan, "Fueling the Fire," 154, describes this mechanism as "the most obvious and widely discussed."

16. Colgan, 154–56; Charles L. Glaser, "How Oil Influences U.S. National Security," *International Security* 38, no. 2 (2013): 123; Georg Strüver and Tim Wegenast, "The Hard Power of Natural Resources: Oil and the Outbreak of Militarized Interstate Disputes," *Foreign Policy Analysis* 14, no. 1 (2016): 86–106.

17. Indra de Soysa, Erik Gartzke, and Tove Grete Lie, "Oil, Blood, and Strategy: How Petroleum Influences Interstate Conflict" (unpublished manuscript, April 15, 2011), 1.

18. For prominent lists, see Francesco Caselli, Massimo Morelli, and Dominic Rohner, "The Geography of Interstate Resource Wars," *Quarterly Journal of Economics* 130, no. 1 (2015): 267–68; Strüver and Wegenast, "Hard Power of Natural Resources"; and Arthur H. Westing, "Appendix 2. Wars and Skirmishes Involving Natural Resources: A Selection from the Twentieth Century," in *Global Resources and International Conflict: Environmental Factors in Strategic Policy and Action*, ed. Arthur H. Westing (Oxford: Oxford University Press, 1986), 204–9.

19. Colgan, "Fueling the Fire," 155; Strüver and Wegenast, "Hard Power of Natural Resources," 103. See also Koubi et al., "Do Natural Resources Matter?," 239.

20. See discussion in chapter 1.

21. Michael T. Klare, *Blood and Oil: The Dangers and Consequences of America's Growing Dependency on Imported Petroleum* (New York: Henry Holt, 2004), 26–55.

22. Bruce Russett, "Security and the Resources Scramble: Will 1984 Be like 1914?," *International Affairs* 58, no. 1 (1981/1982): 48.

23. David G. Victor, "What Resource Wars?," *National Interest* 92 (2007): 48.

1. FROM VALUE TO VIOLENCE

1. This quotation has appeared in *Life*, the *New York Times*, and British parliamentary debates, as well as numerous manuscripts in French and English. In addition, it was quoted as, "One drop of oil is worth one drop of blood of our soldiers," in Hans J. Morgenthau, *Politics among Nations: The Struggle for Power and Peace*, 4th ed. (1948; New York: Alfred A. Knopf, 1967), 111.

2. "Toute défaillance d'essence causerait la paralysie brusque de nos armées et pourrait nous acculer à une paix inacceptable pour les Alliés." All selections from Clémenceau's telegram are from Pierre Fontaine, *La guerre froide du pétrole* (Paris: Editions "Je Sers," 1956), 10.

3. Interestingly, France's petroleum crisis was, itself, an invention. Gregory Nowell, *Mercantile States and the World Oil Cartel, 1900–1939* (Ithaca, NY: Cornell University Press, 1994), 80–81, 108.

4. Erich Zimmermann, *World Resources and Industries*, 2nd rev. ed. (New York: Harper and Brothers, 1951), 814–15. See also Philippe Le Billon, *Wars of Plunder: Conflicts, Profits and the Politics of Resources* (New York: Columbia University Press, 2012), 9–11, 13, 56.

5. Brian C. Black, *Crude Reality: Petroleum in World History*, updated ed. (Lanham, MD: Rowman and Littlefield, 2014), 23; Emma B. Brossard, *Petroleum: Politics and Power* (Tulsa, OK: PennWell, 1983), 1; Dilip Hiro, *Blood of the Earth: The Battle for the World's Vanishing Oil Resources* (New York: Nation Books, 2007), 4; Daniel Yergin, *The Prize: The Epic Quest for Oil, Money, and Power* (New York: Touchstone, 1992), 24.

6. Yergin, *Prize*, 24. See also Black, *Crude Reality*, 22; and Hiro, *Blood of the Earth*, 6.

7. Richard Heinberg, *The Party's Over: Oil, War and the Fate of Industrial Societies* (Gabriola Island, BC: New Society, 2005), 56; Yergin, *Prize*, 24.

8. Black, *Crude Reality*, 17–24.

9. Vaclav Smil, *Oil: A Beginner's Guide* (London: Oneworld, 2008), 92, 101. Wells can now exceed twelve thousand meters in depth.

10. Robert Sollen, *An Ocean of Oil: A Century of Political Struggle over Petroleum off the California Coast* (Juneau, AK: Denali, 1998), xi, 8–9.

11. Smil, *Oil*, 102–3.

12. Those three were the United States, Russia, and Venezuela. *Statistical Yearbook of the League of Nations, 1939/40* (Geneva: League of Nations, 1940), 131.

13. Based on US Energy Information Administration (EIA) data on "crude oil including lease condensate." "International Energy Statistics," EIA, accessed June 21, 2019, https://www.eia.gov/beta/international/data/browser.

14. Black, *Crude Reality*, 25–26.

15. Alison Fleig Frank, *Oil Empire: Visions of Prosperity in Austrian Galicia* (Cambridge, MA: Harvard University Press, 2009), 56–58.

16. The boom was short lived. With the invention of the electric lightbulb in 1880, kerosene demand plummeted. Had it not been for the invention of the automobile, the oil industry would have collapsed.

17. Yergin, *Prize*, 14.

18. The United Kingdom began exploiting North Sea oil in the 1970s.

19. Ian O. Lesser, *Resources and Strategy* (New York: St. Martin's, 1989), 25–26; Peter Tertzakian, *A Thousand Barrels a Second: The Coming Oil Break Point and the Challenges Facing an Energy Dependent World* (New York: McGraw Hill, 2006), 35–38; Yergin, *Prize*, 11–12, 153–56.

20. David A. Deese, "Oil, War, and Grand Strategy," *Orbis* 25, no. 4 (1981): 527–28; Yergin, *Prize*, 167–72.

21. Curzon quoted in Lesser, *Resources and Strategy*, 42–44.

22. The "desert fox" persistently bemoaned the fuel shortages that compelled his retreat. Robert Goralski and Russell W. Freeburg, *Oil and War: How the Deadly Struggle for Fuel in WWII Meant Victory or Defeat* (New York: William Morrow, 1987), 208–17; Yergin, *Prize*, 340–43.

23. Goralski and Freeburg, *Oil and War*, 77, 80–82, 178–80, 183, 250–84; Yergin, *Prize*, 336–38, 346–47.

24. Yergin, *Prize*, 359–61.

25. Yergin, 362–64.

26. For the percentage, see EIA, *Annual Energy Review 2011* (Washington, DC: Office of Energy Statistics, Department of Energy, September 2012), 29. More recent EIA reviews only report energy consumption by source or agency, rather than the two combined. For the number of barrels consumed, see *Fiscal Year 2018 Operational Energy Annual Report* (Washington, DC: Department of Defense, Office of the Under Secretary of Defense for Acquisition, Technology, and Logistics, 2019), 22. This "operational energy" does not include the energy consumed by the DOD's fixed installations ("installation energy"), which accounts for approximately 30 percent of the DOD's energy consumption.

27. *2016 Operational Energy Strategy* (Washington, DC: Department of Defense, 2016), 13–16; *Fiscal Year 2018*, 27–31.

28. EIA, *Monthly Energy Review, July 2017* (Washington, DC: Office of Energy Statistics, Department of Energy, 2017), 37.

29. "Electricity Production from Oil Sources, % of Total," World Bank, accessed June 21, 2019, http://data.worldbank.org/indicator/EG.ELC.PETR.ZS.

30. Black, *Crude Reality*, 163–71; Hiro, *Blood of the Earth*, 33–35.

31. Stephen D. Krasner, *Defending the National Interest: Raw Materials Investments and U.S. Foreign Policy* (Princeton, NJ: Princeton University Press, 1978), 52.

32. Debate continues over the direction of the energy–growth relationship. For a summary, see Ilhan Ozturk, "A Literature Survey on Energy–Growth Nexus," *Energy Policy* 38 (2010): 34–49.

33. James D. Hamilton, "Historical Oil Shocks" (NBER Working Paper No. 16790, National Bureau of Economic Research, Cambridge, MA, 2011); Rebeca Jiménez-Rodríguez and Marcelo Sánchez, "Oil Price Shocks and Real GDP Growth: Empirical Evidence for Some OECD Countries," *Applied Economics* 37, no. 2 (2005): 201–28.

34. Oil-consuming states can also collect substantial revenue by taxing petroleum products, such as gasoline.

35. The share of revenue that a host government receives from oil production depends on numerous factors, including ownership of subsoil resources (public or private), the host state's technological capabilities, the availability of national oil companies, and foreign oil companies' level of interest in the host state's petroleum prospects.

36. "Angola: Analysis," EIA, last updated June 7, 2019, https://www.eia.gov/beta/international/analysis.php?iso=AGO; "Kuwait: Analysis," EIA, last updated November 2, 2016, https://www.eia.gov/beta/international/analysis.php?iso=KWT; "Nigeria: Analysis," EIA, last updated May 6, 2016, https://www.eia.gov/beta/international/analysis.php?iso=NGA.

37. Michael L. Ross, *The Oil Curse: How Petroleum Wealth Shapes the Development of Nations* (Princeton, NJ: Princeton University Press, 2013), 78.

38. Annie Zak, "$1600 PFD Checks Hit Bank Accounts," *Anchorage Daily News*, October 5, 2018.

39. F. Gregory Gause III, *Oil Monarchies: Domestic and Security Challenges in the Arab Gulf States* (New York: Council on Foreign Relations, 1994), 43, 58–66; Ross, *Oil Curse*, 79–80.

40. Michael Penfold-Becerra, "Clientelism and Social Funds: Evidence from Chavez's *Misiones*," *Latin American Politics and Society* 49, no. 4 (2007): 63–84.

41. Stéphan Lacroix, "Comparing the Arab Revolts: Is Saudi Arabia Immune?," *Journal of Democracy* 22, no. 4 (2011): 54.

42. Gause, *Oil Monarchies*, 67–68.

43. Michael Ross, "Will Oil Drown the Arab Spring?," *Foreign Affairs* 90, no. 5 (2011): 4; Sean L. Yom and Gregory Gause III, "Resilient Royals: How Arab Monarchies Hang On," *Journal of Democracy* 23, no. 4 (2012): 74–88.

44. Although see Benjamin Smith, *Hard Times in the Land of Plenty: Oil Politics in Iran and Indonesia* (Ithaca, NY: Cornell University Press, 2007).

45. Ben Hubbard, "Saudi Arabia Restores Public Sector Perks amid Grumbling," *New York Times*, April 23, 2017.

46. Felix Onuah, "Nigeria Resumes Cash Pay-Offs to Former Militants in Oil Hub: Oil Official," Reuters, August 1, 2016.

47. Oil is usually priced in dollars, regardless of where it originates.

48. "Recent Improvements in Petroleum Trade Balance Mitigate U.S. Trade Deficit," EIA, July 21, 2014, https://www.eia.gov/todayinenergy/detail.php?id=17191.

49. It is unclear whether this "oil weapon" can actually compel changes in targeted states' behavior. Roy Licklider, "The Power of Oil: The Arab Oil Weapon and the Netherlands, the United Kingdom, Canada, Japan, and the United States," *International Studies Quarterly* 32, no. 2 (1988): 205–26.

50. For examples, see Rachel Bronson, *Thicker than Oil: America's Uneasy Partnership with Saudi Arabia* (Oxford: Oxford University Press, 2008); and Jeff D. Colgan, "The Emperor Has No Clothes: The Limits of OPEC in the Global Oil Market," *International Organization* 68, no. 3 (2014): 599–632.

51. Hans Morgenthau, *Politics among Nations: The Struggle for Power and Peace*, 7th ed. (1948; New York: McGraw Hill, 2006), 124–25, 128–30, 157; Kenneth N. Waltz, "A Strategy for the Rapid Deployment Force," *International Security* 5, no. 4 (1981): 49–73.

52. There is, of course, a darker side to oil ownership. For a leading discussion of the "resource curse," as applied to oil, see Ross, *Oil Curse.*

53. Indra de Soysa, Erik Gartzke, and Tove Grete Lie, "Oil, Blood, and Strategy: How Petroleum Influences Interstate Conflict" (unpublished manuscript, April 15, 2011), 3.

54. The Biafran War (1967–1970) is a borderline case, as the state of Biafra declared itself independent from Nigeria and was formally recognized by five countries. However, most observers regarded it as a civil war.

55. Oil-related civil wars may involve international actors. However, according to conventional conflict definitions, this does not make them interstate wars.

56. De Soysa, Gartzke, and Lie, "Oil, Blood, and Strategy," 4.

57. This threshold is far lower than the usual figure for interstate wars: one thousand battle deaths. I employ a lower threshold to make the analysis more inclusive.

58. For a list of alternative pathways, see Jeff D. Colgan, "Fueling the Fire: Pathways from Oil to War," *International Security* 38, no. 2 (2013): 156–59.

59. De Soysa, Gartzke, and Lie, "Oil, Blood, and Strategy," 1.

60. Colgan, "Fueling the Fire," 154.

61. Based on EIA data on "crude oil, NGPL, and other liquids" production. "International Energy Statistics."

62. The liberal and realist labels were first used in this debate by Peter Liberman, "The Spoils of Conquest," *International Security* 18, no. 2 (1993): 125–53.

63. Norman Angell, *The Great Illusion: A Study of the Relation of Military Power to National Advantage*, 4th rev. and enlarged ed. (New York: G. P. Putnam's Sons, 1913); Stephen G. Brooks, *Producing Security: Multinational Corporations, Globalization, and the Changing Calculus of Conflict* (Princeton, NJ: Princeton University Press, 2005); Klaus Eugen Knorr, *On the Uses of Military Power in the Nuclear Age* (Princeton, NJ: Princeton University Press, 1966), 73–74; Richard Rosecrance, *The Rise of the Trading State: Commerce and Conquest in the Modern World* (New York: Basic Books, 1986). Brooks does not identify as a liberal, but presents a liberal argument in *Producing Security.*

64. Carl Kaysen, "Is War Obsolete? A Review Essay," *International Security* 14, no. 4 (1990): 54; Knorr, *Uses of Military Power*, 73–74; Rosecrance, *Trading State*, 34. Some realists also make the nationalism argument. See, for example, Stephen Van Evera, *Causes of War: Power and the Roots of Conflict* (Ithaca, NY: Cornell University Press, 1999), 107, 115.

65. Angell, *Great Illusion*, 60–61; Brooks, *Producing Security*, 57–60.

66. Peter Liberman, *Does Conquest Pay? The Exploitation of Occupied Industrial Societies* (Princeton, NJ: Princeton University Press, 1996).

67. John J. Mearsheimer, *The Tragedy of Great Power Politics* (New York: W. W. Norton, 2001), 150.

68. Krasner, *Defending the National Interest*, 337n5.

69. Waltz, "Rapid Deployment Force," 52; Charles L. Glaser, "How Oil Influences U.S. National Security," *International Security* 38, no. 2 (2013): 146.

70. On oil's exceptional importance, see Morgenthau, *Politics* (1967), 111. Morgenthau also added a new section entitled "The Power of Oil" to later editions of *Politics among Nations.*

71. Christopher J. Fettweis, *Dangerous Times? The International Politics of Great Power Peace* (Washington, DC: Georgetown University Press, 2010), 111; Brooks, *Producing Security*, 49. Although he asserts that oil conquest can pay financially, Fettweis also observes that great powers have not fought wars for oil in the Persian Gulf, Caspian Sea, or Pacific Rim. He attributes this outcome to "pacific norms."

72. Daniel Deudney, "Environmental Security: A Critique," in *Contested Grounds: Security and Conflict in the New Environmental Politics*, ed. Daniel Deudney and Richard

Anthony Matthew (Albany: State University of New York Press, 1999), 208; Knorr, *Uses of Military Power*, 24; Richard H. Ullman, *Securing Europe* (Princeton, NJ: Princeton University Press, 1991), 25.

73. Deudney, "Environmental Security," 208.

74. Richard Rosecrance, *The Rise of the Virtual State: Wealth and Power in the Coming Century* (New York: Basic Books, 1999), xiv–xv.

75. Knorr, *Uses of Military Power*, emphasizes oil dependence. Ullman, *Securing Europe*, highlights the lack of oil substitutes and concentration of reserves. Deudney, "Environmental Security," refers to past oil wars.

76. Hanns W. Maull, "Energy and Resources: The Strategic Dimensions," *Survival* 31, no. 6 (1989): 512–13. Bashir argues that cost concerns can also discourage states from milder types of petroleum competition. Omar S. Bashir, "The Great Games Never Played: Explaining Variation in International Competition over Energy," *Journal of Global Security Studies* 2, no. 4 (2017): 288–306.

77. Eugene Gholz and Daryl G. Press, "Protecting 'the Prize': Oil and the U.S. National Interest," *Security Studies* 19, no. 3 (2010): 482. See also Robert H. Johnson, "The Persian Gulf in US Strategy: A Skeptical View," *International Security* 14, no. 1 (1989): 153–54; Thomas L. McNaugher, *Arms and Oil: U.S. Military Strategy and the Persian Gulf* (Washington, DC: Brookings Institution, 1985), 184–95; and Joseph S. Nye, "Energy Nightmares," *Foreign Policy* 40 (Autumn 1980): 143.

78. Kaysen, "Is War Obsolete?," 56–57. See also Rosemary A. Kelanic, "The Petroleum Paradox: Oil, Coercive Vulnerability, and Great Power Behavior," *Security Studies* 25, no. 2 (2016): 181–213.

79. Brenda Shaffer, *Energy Politics* (Philadelphia: University of Pennsylvania Press, 2009), 67.

80. Gholz and Press, "Protecting 'the Prize,'" 482.

81. Daniel Moran and James A. Russell, "Introduction: The Militarization of Energy Security," in *Energy Security and Global Politics: The Militarization of Resource Management*, ed. Daniel Moran and James A. Russell (New York: Routledge, 2009), 5–7.

82. McNaugher, *Arms and Oil*, 185.

83. Evan Luard, *War in International Society: A Study in International Sociology* (London: I. B. Tauris, 1986), 180. Luard does suggest that oil "played some part in motivating Japanese attacks in the Pacific in 1941." But, he states, "it is difficult to point to any other war in which they have had any significant role."

84. Ronnie D. Lipschutz and John P. Holdren, "Crossing Borders: Resource Flows, the Global Environment, and International Security," *Bulletin of Peace Proposals* 21, no. 2 (1990): 123. See also Ronnie D. Lipschutz, *When Nations Clash: Raw Materials, Ideology, and Foreign Policy* (New York: Ballinger, 1989); and Maull, "Energy and Resources," 512.

85. Lesser, *Resources and Strategy*, 183–84.

86. Francesco Caselli, Massimo Morelli, and Dominic Rohner, "The Geography of Interstate Resource Wars," *Quarterly Journal of Economics* 130, no. 1 (2015): 267–315; De Soysa, Gartzke, and Lie, "Oil, Blood, and Strategy"; Erik Gartzke and Dominic Rohner, "Prosperous Pacifists: The Effects of Development on Initiators and Targets of Territorial Conflict" (Working Paper No. 500, Institute for Empirical Research in Economics, University of Zurich, September 2010); Elizabeth Nyman, "Offshore Oil Development and Maritime Conflict in the 20th Century," *Energy Research & Social Science* 6 (2015): 1–7; Kenneth A. Schultz, "Mapping Interstate Territorial Conflict: A New Data Set and Applications," *Journal of Conflict Resolution* 61, no. 7 (2017): 1565–90; Georg Strüver and Tim Wegenast, "The Hard Power of Natural Resources: Oil and the Outbreak of Militarized Interstate Disputes," *Foreign Policy Analysis* 14, no. 1 (2016): 86–106.

87. Caselli, Morelli, and Rohner, "Geography"; Nyman, "Offshore Oil Development"; Strüver and Wegenast, "Hard Power of Natural Resources." Macaulay and Hensel also find that the presence of energy resources (an index variable that includes coal, oil, natural gas, and hydroelectric power) increases the likelihood of territorial claim militarization. Christopher Macaulay and Paul R. Hensel, "Natural Resources and Territorial Conflict" (paper presented at the International Studies Association Annual Convention, Toronto, Canada, March 26–29, 2014).

88. Philippe Le Billon, "Geographies of War: Perspectives on 'Resource Wars,'" *Geography Compass* 1, no. 2 (2007): 163–82. For discussions of these limitations in the quantitative literature on oil and intrastate conflict, see also Nancy Lee Peluso and Michael Watts, eds., *Violent Environments* (Ithaca, NY: Cornell University Press, 2001); Matthew Pritchard, "Re-inserting and Re-politicizing Nature: The Resource Curse and Human-Environment Relations," *Journal of Political Ecology* 20, no. 1 (2013): 361–75; and Frederick Van Der Ploeg and Steven Poelhekke, "The Impact of Natural Resources: Survey of Recent Quantitative Evidence," *Journal of Development Studies* 53, no. 2 (2017): 205–16.

89. The most prominent qualitative studies are Deese, "Oil, War, and Grand Strategy"; Lesser, *Resources and Strategy*; Goralski and Freeburg, *Oil and War*; and Yergin, *Prize*.

90. Cynthia Weber, *International Relations Theory: A Critical Introduction*, 3rd ed. (New York: Routledge, 2010), xxi, 2.

91. Peter M. Haas, "Constructing Environmental Conflicts from Resource Scarcity," *Global Environmental Politics* 2, no. 1 (2002): 1–2.

92. Roger Stern explains the persistence of "oil scarcity ideology" by asserting that it rests on "misperceptions." However, he also asserts that "why misperceptions were so durable deserves more research." Roger J. Stern, "Oil Scarcity Ideology in US Foreign Policy, 1908–97," *Security Studies* 25, no. 2 (2016): 255. My exploration of the oil wars myth provides an answer to Stern's question.

2. EXPLAINING THE OIL WARS MYTH

1. The quotation is from Indra de Soysa, Erik Gartzke, and Tove Grete Lie, "Oil, Blood, and Strategy: How Petroleum Influences Interstate Conflict" (unpublished manuscript, April 15, 2011), 1. Thompson and Rayner define hegemonic myths as "fundamental propositions or assumptions that are unquestionable within the context of a particular discourse." Michael Thompson and Steve Rayner, "Cultural Discourses," in *Human Choice and Climate Change*, ed. Steve Rayner and Elizabeth Malone (Columbus, OH: Battelle, 1998), 1:289. For related discussions of narratives associated with resource scarcity and conflict, as well as why we believe them, see Peter M. Haas, "Constructing Environmental Conflicts from Resource Scarcity," *Global Environmental Politics* 2, no. 1 (2002): 1–2; Elizabeth Hartmann, "Strategic Scarcity: The Origins and Impact of Environmental Conflict Ideas" (PhD diss., London School of Economics and Political Science, 2003); Philippe Le Billon, "Digging into 'Resource War' Beliefs," *Human Geography* 5, no. 2 (2012): 26–40; and Shannon O'Lear, *Environmental Geopolitics* (Lanham, MD: Rowman and Littlefield, 2018).

2. This is similar to a narrative Le Billon identifies in geopolitical theories of resource wars: "resources as loot." Le Billon, "Digging into 'Resource War' Beliefs."

3. For explorations of narratives' ability to shape political beliefs and behaviors, see J. Furman Daniel III and Paul Musgrave, "Synthetic Experiences: How Popular Culture Matters for Images of International Relations," *International Studies Quarterly* 61, no. 3 (2017): 503–16; Ronald R. Krebs, *Narrative and the Making of US National Security* (Cambridge: Cambridge University Press, 2015); and Alexander Spencer, *Romantic Narratives in International Politics: Pirates, Rebels and Mercenaries* (Manchester: Manchester University Press, 2016). For another examination of oil-related narratives that mentions Mad Max,

see Matthew Schneider-Mayerson, *Peak Oil: Apocalyptic Environmentalism and Libertarian Political Culture* (Chicago: University of Chicago Press, 2015).

4. Cynthia Weber, *International Relations Theory: A Critical Introduction*, 3rd ed. (New York: Routledge, 2010), 6.

5. Thomas Hobbes, *Leviathan* (1651; London, Penguin Books, 1985), 185. People who are motivated by the third factor, glory, fight for reputation. This motive, while a plausible incentive for conflict, is less relevant to the oil wars myth.

6. Hobbes, 184–85. For other interpretations of Hobbes's three motives, see Eric J. Hamilton and Brian C. Rathbun, "Scarce Differences: Toward a Material and Systemic Foundation for Offensive and Defensive Realism," *Security Studies* 22, no. 3 (2013): 436–65; John Orme, "The Utility of Force in a World of Scarcity," *International Security* 22, no. 3 (1997/1998): 138–67; and Randall L. Schweller, "Realism and the Present Great Power System: Growth and Positional Conflict over Scarce Resources," in *Unipolar Politics: Realism and State Strategies after the Cold War*, ed. Ethan B. Kapstein and Michael Mastanduno (New York: Columbia University Press, 1999), 28–68.

7. Hans Morgenthau, *Scientific Man vs. Power Politics* (1946; Chicago: University of Chicago Press, 1967), 193. Morgenthau quotes Aristotle on this distinction: "The greatest crimes are caused by excess and not by necessity. Men do not become tyrants in order that they may not suffer cold."

8. Charles L. Glaser, "Political Consequences of Military Strategy: Expanding and Refining the Spiral and Deterrence Models," *World Politics* 44, no. 4 (1992): 501. An alternative way of conceptualizing this dichotomy is to distinguish between "status quo" and "revisionist" states. Randall L. Schweller, "Neorealism's Status-Quo Bias: What Security Dilemma?," *Security Studies* 5, no. 3 (1996): 90–121. Glaser prefers the "not-greedy" label to the "status quo" term, because, he observes, so-called status quo states may feel compelled to expand for security purposes. I replace Glaser's term "not-greedy" with "needy" because it also eliminates that confusion, without resorting to Glaser's self-acknowledged "somewhat awkward terminology" (501).

9. See, for example, Cynthia J. Arnson and I. William Zartman, eds., *Rethinking the Economics of War: The Intersection of Need, Creed, and Greed* (Washington, DC: Woodrow Wilson Center Press, 2005). More often, the distinction is referred to as one between "greed" and "grievance." See Paul Collier and Anke Hoeffler, "Greed and Grievance in Civil War," *Oxford Economic Papers* 56, no. 4 (2004): 563–95; and Mats Berdal and David M. Malone, eds., *Greed and Grievance: Economic Agendas in Civil Wars* (Boulder, CO: Lynne Rienner, 2000).

10. I. William Zartman, "Need, Creed, and Greed in Interstate Conflict," in Arnson and Zartman, *Rethinking the Economics of War*, 263.

11. Cynthia J. Arnson, "The Political Economy of War: Situating the Debate," in Arnson and Zartman, *Rethinking the Economics of War*, 11.

12. George Miller, dir., *Mad Max* (South Yarra, Australia: Roadshow Film Distributors, 1979).

13. George Miller, dir., *Mad Max 2: The Road Warrior* (Australia: Warner Bros. Pictures, 1981).

14. *Road Warrior.*

15. Thomas Robert Malthus, *On Population*, ed. Gertrude Himmelfarb (1798; New York: Random House, 1960), 9, 57.

16. Malthus, 20–21.

17. Malthus, 175.

18. Thomas Robert Malthus, *An Essay on the Principle of Population*, vol. 1 (1798; New York: Cosimo Classics, 2007).

19. Charles Darwin, *The Autobiography of Charles Darwin, 1809–1882* (1887; New York: W. W. Norton, 1958), 120.

20. Charles Darwin, "Letter 71, to A. R. Wallace, Down, 6 April 1859," in *More Letters of Charles Darwin: A Record of His Work in a Series of Hitherto Unpublished Letters*, ed. Francis Darwin and Albert Charles Seward (London: John Murray, 1903), 1:294.

21. Charles Darwin, "Essay of 1844," in *Evolution by Natural Selection*, by Charles Darwin and Alfred Russel Wallace (New York: Cambridge University Press, 1871; New York: Johnson Reprint, 1958), 116. Citations refer to the 1958 edition.

22. Charles Darwin, *On the Origin of Species by Means of Natural Selection, or, The Preservation of Favoured Races in the Struggle for Life* (London: Grant Richards, 1902), 4, 59.

23. Darwin, "Essay of 1844," 116–17.

24. James Allan Rogers, "Darwinism and Social Darwinism," *Journal of the History of Ideas* 33, no. 2 (1972): 275.

25. Darwin later adopted the phrase "survival of the fittest," stating that it was equivalent to natural selection. Rogers, 277–78.

26. There was a revival of this type of thinking in the 2000s. See, for example, Bradley A. Thayer, *Darwin and International Relations: On the Evolutionary Origins of War and Ethnic Conflict* (Lexington: University Press of Kentucky, 2004); and William R. Thompson, ed., *Evolutionary Interpretations of World Politics* (New York: Routledge, 2001).

27. Christian Abrahamsson, "On the Genealogy of Lebensraum," *Geographica Helvetica* 68, no. 1 (2013): 39–40; Woodruff D. Smith, "Friedrich Ratzel and the Origins of Lebensraum," *German Studies Review* 3, no. 1 (1980): 53–54. The term *lebensraum* was coined by a German biologist reviewing Darwin's *Origin of Species*.

28. David Atkinson, "Geopolitical Imaginations in Modern Italy," in *Geopolitical Traditions: A Century of Geopolitical Thought*, ed. Klaus Dodds and David Atkinson (London: Routledge, 2000), 102–4; David Thomas Murphy, *The Heroic Earth: Geopolitical Thought in Weimar Germany, 1918–1933* (Kent, OH: Kent State University Press, 1997), 35, 96, 138, 194–96, 204.

29. Vladimir I. Lenin, *Imperialism: The Highest Stage of Capitalism* (1917; New York: International, 2002).

30. German geopolitical theories in the interwar period are most strongly associated with Karl Haushofer, whose thinking purportedly influenced Adolf Hitler.

31. Nazli Choucri and Robert C. North, "Dynamics of International Conflict: Some Policy Implications of Population, Resources, and Technology," *World Politics* 24, no. 1 (1972): 86.

32. Choucri and North, 94–95, 105–8. See also Nazli Choucri and Robert C. North, *Nations in Conflict: National Growth and International Violence* (San Francisco: W. H. Freeman, 1975).

33. Paul R. Ehrlich, *The Population Bomb* (New York: Ballantine Books, 1968).

34. Donella H. Meadows et al., *The Limits to Growth: A Report for the Club of Rome's Project on the Predicament of Mankind* (New York: New American Library, 1972).

35. Stewart Udall, "The Last Traffic Jam," *Atlantic Monthly*, October 1972.

36. James Akins, "The Oil Crisis: This Time, the Wolf Is Here," *Foreign Affairs* 51, no. 3 (April 1973): 462–490.

37. Meg Jacobs, *Panic at the Pump: The Energy Crisis and the Transformation of American Politics in the 1970s* (New York: Farrar, Straus and Giroux, 2016).

38. Michael J. Graetz, *The End of Energy: The Unmaking of America's Environment, Security, and Independence* (Cambridge, MA: MIT Press, 2011).

39. Henry Kissinger, interview by *Business Week*, December 23, 1974, published in the *Department of State Bulletin*, January 27, 1975, 101.

40. Jock A. Finlayson and David G. Haglund, "Whatever Happened to the Resource War?," *Survival* 29, no. 5 (1987): 403; Haig quoted in Ian O. Lesser, *Resources and Strategy* (New York: St. Martin's, 1989), 153.

41. For example, see William J. Broad, "Resource Wars: The Lure of South Africa," *Science* 210, no. 4474 (1980): 1099–100; Geoffrey Kemp, "Scarcity and Strategy," *Foreign Affairs* 56, no. 2 (January 1978): 396–414; Bruce Russett, "Security and the Resources Scramble: Will 1984 Be like 1914?," *International Affairs* 58, no. 1 (1981/1982): 42–58; and Michael Tanzer, *The Race for Resources: Continuing Struggles over Minerals and Fuels* (New York: Monthly Review Press, 1980).

42. Finlayson and Haglund, "Whatever Happened?"; Bruce Russett, "Dimensions of Resource Dependence: Some Elements of Rigor in Concept and Policy Analysis," *International Organization* 38, no. 3 (1984): 481–99.

43. Thomas Homer-Dixon, "Environmental Scarcities and Violent Conflict: Evidence from Cases," *International Security* 19, no. 1 (1994): 5–40.

44. Robert D. Kaplan, "The Coming Anarchy," *Atlantic Monthly*, February 1994.

45. Phillip Stalley, "Environmental Scarcity and International Conflict," *Conflict Management and Peace Science* 20, no. 2 (2003): 35.

46. William J. Clinton, "Remarks to the National Academy of Sciences," June 29, 1994, in *Public Papers of the Presidents of the United States: William J. Clinton (1994, Book 1)*, January 1–July 31, 1994 (Washington, DC: Office of the Federal Register National Archives and Record Administration), 1162.

47. Stalley, "Environmental Scarcity," 35.

48. Michael T. Klare, *Rising Powers, Shrinking Planet: The New Geopolitics of Energy* (New York: Henry Holt, 2008), 20, 147, 149. Ciută asserts that a "quasi-Darwinian logic of survival" characterizes one of three "logics" of energy security: the "logic of war." Felix Ciută, "Conceptual Notes on Energy Security: Total or Banal Security?," *Security Dialogue* 41, no. 2 (2010): 130.

49. Michael T. Klare, *The Race for What's Left: The Global Scramble for the World's Last Resources* (New York: Henry Holt, 2012), 214–15.

50. Klare, 210.

51. Klare, *Rising Powers*, 30.

52. Michael T. Klare, *Resource Wars: The New Landscape of Global Conflict* (New York: Henry Holt, 2001), 14.

53. For example, see Nader Elhefnawy, "The Impending Oil Shock," *Survival: Global Politics and Strategy* 50, no. 2 (2008): 37–66; Jörg Friedrichs, "Global Energy Crunch: How Different Parts of the World Would React to a Peak Oil Scenario," *Energy Policy* 38 (2010): 4562–69; Dambisa Moyo, *Winner Take All: China's Race for Resources and What It Means for the World* (New York: Basic Books, 2012), 45–74; Susanne Peters, "Coercive Western Energy Security Strategies: 'Resource Wars' as a New Threat to Global Security," *Geopolitics* 9, no. 1 (2004): 187–212; and Mamdouh G. Salameh, "Quest for Middle East Oil: The US versus the Asia-Pacific Region," *Energy Policy* 31, no. 11 (2003): 1085–91.

54. Robert N. Bradbury, dir., *Riders of Destiny* (Los Angeles: Monogram Pictures, 1933); Joseph Kane, dir., *King of the Pecos* (Los Angeles: Republic Pictures, 1936).

55. William Wyler, dir., *The Big Country* (Beverly Hills, CA: United Artists, 1958). Other Westerns involving water conflicts include Spencer Gordon Bennet, dir., *Law of the Ranger* (Culver City, CA: Columbia Pictures, 1937); Lesley Selander, dir., *Stampede* (Glendale, CA: Allied Artists Pictures, 1949); and Clint Eastwood, dir., *Pale Rider* (Burbank, CA: Warner Bros., 1985).

56. Roman Polanski, dir., *Chinatown* (Los Angeles: Paramount Pictures, 1974); Robert Redford, dir., *The Milagro Beanfield War* (Universal City, CA: Universal Pictures, 1988).

57. Rachel Talalay, dir., *Tank Girl* (Beverly Hills, CA: United Artists, 1995); Marc Forster, dir., *Quantum of Solace* (Culver City, CA: Columbia Pictures, 2008); George Miller, dir., *Mad Max: Fury Road* (Burbank, CA: Warner Bros., 2015).

58. The antagonists in these films, in contrast, are often motivated by greed.

59. James Cameron, dir., *Avatar* (Los Angeles, CA: Twentieth Century Fox, 2009); Neill Blomkamp, dir., *Elysium* (Culver City, CA: Sony Pictures Releasing, 2013); Christopher Nolan, dir., *Interstellar* (Los Angeles: Paramount Pictures, 2014); Andrew Stanton, dir., *Wall-E* (Burbank, CA: Walt Disney Studios Motion Pictures, 2008).

60. Michael Anderson, dir., *Logan's Run* (Beverly Hills, CA: United Artists, 1976); Richard Fleischer, dir., *Soylent Green* (Beverly Hills, CA: Metro-Goldwyn-Mayer, 1973).

61. *The Walking Dead*, created by Frank Darabont and Angela Kang, originally aired on American Movie Classics (AMC) (2010–), distributed in the United States by Netflix; *Into the Badlands*, created by Alfred Gough and Miles Millar, originally aired on American Movie Classics (AMC) (2015–2019) distributed in the United States by Netflix.

62. Byron Haskin, dir., *The War of the Worlds* (Los Angeles, CA : Paramount Pictures, 1953).

63. Roland Emmerich, dir., *Independence Day* (Los Angeles, CA: Twentieth Century Fox, 1996); Jonathan Liebesman, dir., *Battle: Los Angeles* (Culver City, CA: Columbia Pictures, 2011); Joseph Kosinski, dir., *Oblivion* (Universal City, CA: Universal Pictures, 2013).

64. *Occupied*, created by Karianne Lund, Erik Skjoldbjærg, and Jo Nesbø, originally aired as *Okkupert* (2015–), TV2 Norway, distributed in the United States by Netflix.

65. Sydney Pollack, dir., *Three Days of the Condor* (Hollywood, CA: Paramount Pictures, 1975).

66. Massimo Livi Bacci, *El Dorado in the Marshes: Gold, Slaves and Souls between the Andes and the Amazon*, trans. Carl Ipsen (Cambridge, UK: Polity, 2010), 14.

67. Herbert E. Bolton, *Coronado: Knight of Pueblos and Plains* (Albuquerque: University of New Mexico Press, 1990), 4. Originally published in 1949 as *Coronado on the Turquoise Trail, Knight of Pueblos and Plains.*

68. Charles Nicholl, *The Creature in the Map: A Journey to El Dorado* (Chicago: University of Chicago Press, 1997), 9.

69. Adolph Francis Alphonse Bandelier, *The Gilded Man: (El Dorado) and Other Pictures of the Spanish Occupancy of America* (New York: D. Appleton, 1893); Shannon L. Kenny, *Gold: A Cultural Encyclopedia* (Santa Barbara, CA: ABC-CLIO, 2011), 92; Livi Bacci, *El Dorado in the Marshes*, 13, 17.

70. Walker Chapman, *The Golden Dream: Seekers of El Dorado* (Indianapolis: Bobbs-Merrill, 1967), 30–51.

71. Nicholl, *Creature in the Map.*

72. Chapman, *Golden Dream*, 32.

73. On the greed these expeditions entailed, see John Silver, "The Myth of El Dorado," *History Workshop Journal*, no. 34 (1992): 3–4.

74. Frank Merchant, "Legend and Fact about Gold in Early America," *Western Folklore* 13, no. 2/3 (1954): 172–74; Ralph H. Vigil, "Spanish Exploration and the Great Plains in the Age of Discovery: Myth and Reality," *Great Plains Quarterly* 10, no. 1 (1990), 9–10. The Seven Cities were originally depicted in a medieval legend about seven bishops fleeing the eighth-century Moorish conquest of Portugal to found seven cities on the island of "Antilia," off the Iberian coast. The sand on the islands was purportedly one-third gold. William H. Babcock, "The Island of the Seven Cities," *Geographical Review* 7, no. 2 (1919): 98–106. This story shifted to the American Southwest in the 1530s, after the Spanish heard tales of immensely wealthy settlements in the area. Bolton, *Coronado*, 6, 27–31.

75. Richard W. Berber, *The Holy Grail: Imagination and Belief* (Cambridge, MA: Harvard University Press, 2004).

76. For a prominent discussion of this mechanism, see Michael Ross, "A Closer Look at Oil, Diamonds, and Civil War," *Annual Review of Political Science* 9 (2006): 265–300.

77. Jack Snyder, "Imperial Temptations," *National Interest* 71 (2003): 35–36.

78. Gartzke and Rohner assert that Saddam invaded because he believed it would be "profitable." Erik Gartzke and Dominic Rohner, "To Conquer or Compel: War, Peace, and Economic Development" (Working Paper No. 511, Institute for Empirical Research in Economics, University of Zurich, September 2010), 7. See also Daniel Yergin, *The Prize: The Epic Quest for Oil, Money, and Power* (New York: Touchstone, 1992), 12.

79. Ovid, *Metamorphoses*, trans. John Dryden and others, ed. Sir Samuel Garth (1717; London: Wordsworth Editions Limited, 1998), 7–8.

80. Ovid, 7.

81. *Virgil's Aeneid*, trans. John Dryden (New York: P. F. Collier and Son, 1909), 87.

82. *Virgil's Aeneid*, 133.

83. Joshua 7:126.

84. Dante Alighieri, *The Convivio of Dante Alighieri* (London: J. M. Dent, 1903), 287.

85. Dante, 286–87.

86. Geoffrey Chaucer, *The Canterbury Tales*, trans. J. U. Nicholson (1934; Mineola, NY: Dover, 2004), 255–67.

87. John Huston, dir., *The Treasure of the Sierra Madre* (Burbank, CA: Warner Bros., 1948).

88. Huston.

89. Bibo Bergeron and Don Paul, dirs., *The Road to El Dorado* (Universal City, CA: Dreamworks Distribution, 2000).

90. Steven Spielberg, dir., *Indiana Jones and the Kingdom of the Crystal Skull* (Los Angeles, CA: Paramount Pictures, 2008); Steven Spielberg, dir., *Indiana Jones and the Last Crusade* (Los Angeles, CA: Paramount Pictures, 1989).

91. Jon Turteltaub, dir., *National Treasure: Book of Secrets* (Burbank, CA: Walt Disney Studios Motion Pictures, 2007).

92. Byron Haskin, dir., *Treasure Island* (Santa Monica, CA: RKO Radio Pictures, 1950); Sergio Leone, dir., *The Good, the Bad, and the Ugly* (Beverly Hills, CA: United Artists, 1967); David O. Russell, dir., *Three Kings* (Burbank, CA: Warner Bros., 1999); Richard Donner, dir., *Goonies* (Burbank, CA: Warner Bros., 1985); Gore Verbinski, dir., *The Pirates of the Caribbean: The Curse of the Black Pearl* (Burbank, CA: Buena Vista Pictures, 2003).

93. Charles Rooking Carter, *Victoria, the British "El Dorado": Or: Melbourne in 1869; Shewing the Advantages of That Colony as a Field for Emigration* (London: Edward Stanford, 1870); J. S. Holliday, *The World Rushed In: The California Gold Rush Experience* (1981; Norman: University of Oklahoma Press, 2015), 9, 49; Edward Peter Mathers, *Zambesia, England's El Dorado in Africa: An Account of the Gold Fields of British South Africa* (London: King and Sell, 1891); Catherine Holder Spude et al., eds., *Eldorado! The Archaeology of Gold Mining in the Far North* (Lincoln: University of Nebraska Press, 2011).

94. G. Gregory Crampton, "The Myth of El Dorado," *Historian* 13, no. 2 (1951): 127–28.

95. Dwight Swain, foreword to *The Call of the Wild*, by Jack London (New York: Tor Classics, 1990), viii, ix; Ralph P. Bieber, "California Gold Mania," *Mississippi Valley Historical Review* 35, no. 1 (1948): 3–28.

96. Connelley quoted in Holliday, *The World Rushed In*, 507.

97. Malcolm J. Rohrbough, *Days of Gold: The California Gold Rush and the American Nation* (Berkeley: University of California Press, 1997), 87, 127, 218.

98. Hildegarde Dolson, *The Great Oildorado* (New York: Random House, 1959).

99. William Rintoul, *Oildorado: Boom Times on the West Side* (Fresno, CA: Valley, 1978).

100. El Dorado Drilling Company, website, accessed March 10, 2018, http://www.eldoradodrillingcompany.net; Dorado Oil Company, website, accessed July 15, 2019, http://www.doradooil.com. Over the course of this book's production, another company website—for El Dorado Oil and Gas, Inc.—has disappeared, possibly illustrating the risks associated with oil pursuits all too well.

101. George Stevens, dir., *Giant* (Burbank, CA: Warner Bros., 1956).

102. Michael Watts, "Petro-violence: Community, Extraction, and Political Ecology of a Mythic Commodity," in *Violent Environments*, ed. Nancy Lee Peluso and Michael Watts (Ithaca, NY: Cornell University Press, 2001), 190.

103. Terry Lynn Karl, *The Paradox of Plenty: Oil Booms and Petro-states* (Berkeley: University of California Press, 1997), 23–43.

104. Leonardo Maugeri, *The Age of Oil: The Mythology, History, and Future of the World's Most Controversial Resource* (Westport, CT: Praeger, 2006).

105. Watts, "Petro-violence," 191.

106. Timothy C. Winegard, *The First World Oil War* (Toronto: University of Toronto Press, 2016), 7.

107. Ryszard Kapuscinski, *Shah of Shahs*, trans. William R. Brand and Katarzyna Mroczkowska-Brand (London: Quartet Books, 1985), 34–35.

108. Kapuscinski, 35.

109. Brian Black, *Petrolia: The Landscape of America's First Oil Boom* (Baltimore: Johns Hopkins University Press, 2000), 71.

110. Paul Thomas Anderson, dir., *There Will Be Blood* (Los Angeles, CA: Paramount Vantage, 2007).

111. Upton Sinclair, *Oil! A Novel* (1927; New York: Penguin Books, 2007), 160. For another exploration of oil in film, see Robert Lifset and Brian C. Black, "Imaging the 'Devil's Excrement': Big Oil in Petroleum Cinema, 1940–2007," *Journal of American History* 99, no. 1 (2012): 135–44.

112. Sinclair, *Oil!*, 489.

113. Anderson, *There Will Be Blood.*

3. WHY CLASSIC OIL WARS DO NOT PAY

1. This study examines only the oil-related obstacles to classic oil wars. There are, of course, other impediments to classic oil wars, including the costs of the manpower and materiel required to prosecute the conflicts, the damage that oil-related contention can cause to belligerents' broader political and economic relations, and the opportunity costs of conflict. If all of these obstacles are taken into account, then fighting for oil never pays. However, my argument aims to demonstrate that fighting for oil has limited utility, even if only oil-related obstacles are taken into account. In focusing on the variable of interest—oil—rather than wars' total cost, the study mirrors earlier contributions to the value of conquest debate, such as Stephen G. Brooks, *Producing Security: Multinational Corporations, Globalization, and the Changing Calculus of Conflict* (Princeton, NJ: Princeton University Press, 2005), 161; and Peter Liberman, *Does Conquest Pay? The Exploitation of Occupied Industrial Societies* (Princeton, NJ: Princeton University Press, 1996), x.

2. The phrase is from Carl Kaysen, "Is War Obsolete? A Review Essay," *International Security* 14, no. 4 (1990): 54. But he does not apply it to oil wars.

3. On the lootability of primary commodities, see Norman Angell, *The Great Illusion: A Study of the Relation of Military Power to National Advantage*, 4th rev. and enlarged ed. (New York: G. P. Putnam's Sons, 1913), 51, 108–9; and Klaus Eugen Knorr, *On the Uses of Military Power in the Nuclear Age* (Princeton, NJ: Princeton University Press, 1966), 21.

4. Quoted in Liberman, *Does Conquest Pay?*, 8.

5. Philippe Le Billon, *Wars of Plunder: Conflicts, Profits and the Politics of Resources* (New York: Columbia University Press, 2012), 73.

6. This figure quantifies the accuracy of air strikes against Germany's synthetic materials industries. Presumably, strikes on other targets were equally inaccurate. US Strategic Bombing Survey, *Oil Division: Final Report* (Washington, DC: US Government Printing Office, January 1947), figure 7.

7. Ronald E. Bergquist, *The Role of Airpower in the Iran-Iraq War* (Washington, DC: US Government Printing Office, 1988), 44, 47; John Bulloch and Harvey Morris, *The Gulf War: Its Origins, History, and Consequences* (London: Methuen, 1989), 40–41; Phebe Marr, *The Modern History of Iraq* (Boulder, CO: Westview, 1985), 294.

8. Abbas Alnasrawi, "Economic Consequences of the Iran–Iraq War," *Third World Quarterly* 8, no. 3 (1986): 873. Iran's output dropped from 1.3 million barrels per day to 450,000 barrels per day.

9. Bergquist, *Role of Airpower*, 45–46; Bulloch and Morris, *Gulf War*, 41; Marr, *Modern History of Iraq*, 300; Kamran Mofid, *The Economic Consequences of the Gulf War* (London: Routledge, 1990), 38.

10. Alnasrawi, "Economic Consequences," 873. Iraq's output dropped from 3.4 million barrels per day to 140,000 barrels per day.

11. The phrase is from Stephen Van Evera, *Causes of War: Power and the Roots of Conflict* (Ithaca, NY: Cornell University Press, 1999), 112.

12. Robert Goralski and Russell W. Freeburg, *Oil and War: How the Deadly Struggle for Fuel in WWII Meant Victory or Defeat* (New York: William Morrow, 1987), 141–42, 182; Daniel Yergin, *The Prize: The Epic Quest for Oil, Money, and Power* (New York: Touchstone, 1992), 181, 337, 351–53.

13. Irvine H. Anderson Jr., *The Standard-Vacuum Oil Company and United States East Asian Policy, 1933–1941* (Princeton, NJ: Princeton University Press, 1975), 193.

14. Bob Tippee, "Kuwait Pressing towards Preinvasion Oil Production Capacity," *Oil and Gas Journal*, March 5, 1993.

15. Tippee; Jerome B. Cohen, *Japan's Economy in War and Reconstruction* (Minneapolis: University of Minnesota Press, 1949), 140; Yergin, *Prize*, 181, 351–53.

16. Joshua R. Itzkowitz Shifrinson and Miranda Priebe, "A Crude Threat: The Limits of an Iranian Missile Campaign against Saudi Arabian Oil," *International Security* 36, no. 1 (2011): 167–201.

17. "Nigerian Attack Closes Oilfield," *BBC News*, June 20, 2008, http://news.bbc.co.uk/2/hi/africa/7463288.stm

18. Angell, *Great Illusion*, 135; Stephen D. Krasner, *Defending the National Interest: Raw Materials Investments and U.S. Foreign Policy* (Princeton, NJ: Princeton University Press, 1978), 275; Richard Rosecrance, *The Rise of the Trading State: Commerce and Conquest in the Modern World* (New York: Basic Books, 1986), 32–35; Richard Rosecrance, *The Rise of the Virtual State: Wealth and Power in the Coming Century* (New York: Basic Books, 1999), 7; Van Evera, *Causes of War*, 107, 115.

19. Knorr, *Uses of Military Power*, 74.

20. Krasner, *Defending the National Interest*, 275.

21. For example, Knorr, *Uses of Military Power*, 74; Liberman, *Does Conquest Pay?*, 19; and Rosecrance, *Trading State*, 34. Liberman uses the term *pre-industrial* to refer to these states.

22. Critics of oil war arguments have highlighted this obstacle. See, for example, David R. Henderson, "Do Governments Need to Go to War for Oil?," in *Handbook of Oil Politics*, ed. Robert E. Looney (New York: Routledge, 2012), 142–43.

23. Anthony Sampson, *The Seven Sisters: The Great Oil Companies and the World They Shaped* (1975; New York: Bantam Books, 1991).

24. Prominent nationalizations in the 1960s included Iraq (1961), Burma (1962), Egypt (1962), Argentina (1963), Indonesia (1963), and Peru (1968). Iran had nationalized its oil in 1951.

25. Randal C. Archibold and Elisabeth Malkin, "Mexico's Pride, Oil, May Be Opened to Outsiders," *New York Times*, December 12, 2013.

26. Parisa Hafezi and Rania El Gamal, "Cancellation of Iran Oil Contracts' Presentation Signals Infighting," Reuters, February 11, 2016.

27. "Saudi Aramco IPO Sparks Fears of Loss of Cash Cow," *Financial Times*, May 9, 2017.

28. "'Bin Laden' Tape Urges Oil Attack," *BBC News*, December 16, 2004, http://news.bbc.co.uk/2/hi/middle_east/4101021.stm.

29. Based on data from the Energy Infrastructure Attack Database. Jennifer Giroux, Peter Burgherr, and Laura Melkunaite, "Research Note on the Energy Infrastructure Attack Database (EIAD)," *Perspectives on Terrorism* 7, no. 6 (2013): 113–25.

30. "ABB Lummus Vacates Yanbu; Saudis Vow to Crush Terrorists," *Oil & Gas Journal*, May 10, 2004; "Gunmen 'Killed 22' in Saudi City," *BBC News*, May 30, 2004, http://news.bbc.co.uk/2/hi/middle_east/3762423.stm.

31. Michael L. Ross, *The Oil Curse: How Petroleum Wealth Shapes the Development of Nations* (Princeton, NJ: Princeton University Press, 2013), 173–74. These kidnapping incidents increased after the route was completed, as rebels had discovered that ransom payments were an extremely lucrative funding source. See also Le Billon, *Wars of Plunder*, 74–75.

32. For examples, see Michael Watts, "Petro-insurgency or Criminal Syndicate? Conflict and Violence in the Niger Delta," *Review of African Political Economy* 34, no. 114 (2007): 637–60.

33. Daniel Moran and James A. Russell, "Introduction: The Militarization of Energy Security," in *Energy Security and Global Politics: The Militarization of Resource Management*, ed. Daniel Moran and James A. Russell (New York: Routledge, 2009), 11.

34. Jennifer Giroux, "A Portrait of Complexity: New Actors and Contemporary Challenges in the Global Energy System and the Role of Energy Infrastructure Security," *Risk, Hazards and Crisis in Public Policy* 1, no. 1 (2010): 46.

35. Eugene Gholz and Daryl G. Press, "Footprints in the Sand," *American Interest*, March/April 2010, 62.

36. Khalid R. Al-Rodhan, *The Impact of the Abqaiq Attack on Saudi Energy Security* (Washington, DC: Center for Strategic and International Studies, February 2006). On the broader issue of Saudi infrastructure security, see Anthony H. Cordesman and Nawaf Obaid, *Saudi Petroleum Security: Challenges and Responses* (Washington, DC: Center for Strategic and International Studies, November 2004).

37. Hugh Naylor, "The Islamic State Threatens to Capture Massive Iraqi Oil Refinery," *Washington Post*, May 15, 2015.

38. In their analysis of Abqaiq's vulnerability, Shifrinson and Priebe, "Crude Threat," conclude that rendering the center inoperable would be extremely challenging for a foreign military (184–87). However, they also concede that local insurgent groups might be able to launch more effective attacks against these types of targets (199).

39. Al-Rodhan, *Impact*, 4–5.

40. Shifrinson and Priebe, "Crude Threat," 177.

41. Elisha Bala-Gbogbo, "Shell to Chevron Move Offshore as Nigerian Risks Mount," *Bloomberg*, July 31, 2013.

42. Giroux, Burgherr, and Melkunaite, "Research Note," 120.

43. Tim Pippard, "'Oil-Qaeda': Jihadist Threats to the Energy Sector," *Perspectives on Terrorism* 4, no. 3 (2010): 3–14.

44. Kathryn Nwajiaku-Dahou, *The Politics of Amnesty in the Niger Delta: Challenges Ahead* (Paris: Institut Français des Relations Internationales, December 2010).

45. Maggie Flick, "Nigeria Oil Infrastructure Threatened by Cuts to Amnesty Fund," *Financial Times*, May 9, 2016; Felix Onuah, "Nigeria Resumes Cash Pay-Offs to Former Militants in Oil Hub: Oil Official," Reuters, August 1, 2016; "Crude Oil Disruptions in Nigeria Increase as a Result of Militant Attacks," Today in Energy series, US Energy Information Administration, August 18, 2016, https://www.eia.gov/todayinenergy/detail.php?id=27572.

46. Knorr, *Uses of Military Power*, 74.

47. Gholz and Press, "Footprints," 62.

48. Eugene Gholz and Daryl G. Press, "Protecting 'the Prize': Oil and the U.S. National Interest," *Security Studies* 19, no. 3 (2010): 464, 470.

49. Fred Burton, "Saudi Arabia: The Shift toward Precision Strikes," *STRATFOR Global Intelligence*, June 4, 2004.

50. Liberman, *Does Conquest Pay?*

51. Gholz and Press, "Protecting 'the Prize,'" 482.

52. Glenn Frankely, "U.S. Mulled Seizing Oil Fields in '73," *Washington Post*, January 1, 2004. For the classic discussion of this episode, see Thomas L. McNaugher, *Arms and Oil: U.S. Military Strategy and the Persian Gulf* (Washington, DC: Brookings Institution, 1985), 184–95.

53. Drew Middleton, "Military Men Challenge Mideast 'Force' Strategy," *New York Times*, January 10, 1975.

54. Knorr, *Uses of Military Power*, 12, 67–68; Van Evera, *Causes of War*, 112, 116. Realists and liberals make this balancing argument.

55. Tanisha M. Fazal, *State Death: The Politics and Geography of Conquest, Occupation, and Annexation* (Princeton, NJ: Princeton University Press, 2007), 44–54. The norm against conquest is similar to the "territorial integrity norm," which maintains that states should not attempt to alter international boundaries by force. Mark W. Zacher, "The Territorial Integrity Norm: International Boundaries and the Use of Force," *International Organization* 55, no. 2 (2001): 215–50.

56. M. Taylor Fravel, "International Relations Theory and China's Rise: Assessing China's Potential for Territorial Expansion," *International Studies Review* 12, no. 4 (2010): 511. Fravel is specifically referring to the international community's normative incentive to respond to Iraq's 1990 invasion of Kuwait.

57. Jeff D. Colgan, "Fueling the Fire: Pathways from Oil to War," *International Security* 38, no. 2 (2013): 156–59; Indra de Soysa, Erik Gartzke, and Tove Grete Lie, "Oil, Blood, and Strategy: How Petroleum Influences Interstate Conflict" (unpublished manuscript, April 15, 2011), 9–16; Gholz and Press, "Protecting 'the Prize,'" 455.

58. President George H. W. Bush, speech to Congress, September 11, 1990, http://millercenter.org/the-presidency/presidential-speeches/september-11-1990-address-joint-session-congress.

59. "Russia Plants Flag under N. Pole," *BBC News*, August 2, 2007, http://news.bbc.co.uk/2/hi/europe/6927395.stm

60. Leigh Ann Caldwell, "Trump Said 'Take the Oil' from Iraq. Can He?," *NBC News*, September 8, 2016, https://www.nbcnews.com/politics/2016-election/trump-said-take-oil-can-he-n645021; Steven Mufson, "Trump's Illegal, Impossible, and 'beyond Goofy' Idea of Seizing Iraq's Oil," *Washington Post*, September 9, 2016; Damien Paletta, "Trump's 'Take the Oil' Plan Would Violate Geneva Conventions, Experts Say," *Wall Street Journal*, September 8, 2016.

61. United Nations, "Charter of Economic Rights and Duties of States" (1974), art. 2, para. 1.

62. "Convention (IV) Respecting the Laws and Customs of War on Land and Its Annex: Regulations concerning the Laws and Customs of War on Land," The Hague (1907), arts. 46, 55.

63. Kenneth N. Waltz, "A Strategy for the Rapid Deployment Force," *International Security* 5, no. 4 (1981): 62.

64. Middleton, "Military Men."

65. There are extensive, ongoing debates about the efficacy of international sanctions. For two prominent contributions, see Daniel W. Drezner, "The Hidden Hand of Economic Coercion," *International Organization* 57, no. 3 (2003): 643–59; and Robert A. Pape, "Why Economic Sanctions Do Not Work," *International Security* 22, no. 2 (1997): 90–136.

66. Based on export data from the US Energy Information Administration. Exports dropped from 2.47 million barrels per day in 1989 to 42,000 barrels per day in 1991. "International Energy Statistics," US Energy Information Administration, accessed June 21, 2019, https://www.eia.gov/beta/international/data/browser.

67. UN Resolution 986 created the oil-for-food program in April 1995. However, phase 1 was not implemented until December 1996.

68. Yergin, *Prize*, 347. They also struck German synthetic fuel facilities.

69. Thomas A. Keaney and Eliot A. Cohen, *Revolution in Warfare? Air Power in the Persian Gulf* (Annapolis, MD: Naval Institute Press, 1995), 11–14, 61–66.

70. Congressional Research Service, *Oil Fields as Military Objectives: A Feasibility Study*, 94th Cong., 1st Sess., prepared for the Special Subcommittee on Investigations of the Committee on International Relations (Washington, DC: US Government Printing Office, August 21, 1975).

71. Patrick Armstrong and Vivian Forbes, *The Falkland Islands and Their Adjacent Maritime Area*, Maritime Briefing 2, no. 3 (Durham, UK: International Boundaries Research Unit, 1997), 26–28; Bruce Blanche, "The Falkland Islands: The New Hydrocarbon Frontier," in *Boundaries and Energy: Problems and Prospects*, ed. Gerald Henry Blake (The Hague: Kluwer Law International, 1998), 334.

72. Angell, *Great Illusion*, 63, 127–28.

73. Brooks, *Producing Security*, 59.

74. Aggressors are effectively "roving bandits." Mancur Olson, "Dictatorship, Democracy, and Development," *American Political Science Review* 87, no. 3 (1993): 567–76.

75. Brooks, *Producing Security*, 59. See also Charles Lipson, *Standing Guard: Protecting Foreign Capital in the Nineteenth and Twentieth Centuries* (Berkeley: University of California Press, 1985).

76. Angell, *Great Illusion*, 128.

77. Brooks, *Producing Security*, 59.

78. Between 2011 and 2012, Rosneft signed agreements with BP, Eni, Exxon, and Statoil for development in the Arctic. "Statoil Clinches Deal with Rosneft on Exploration in Russian Arctic and Sea of Okhotsk," *IHS World Markets Energy Perspective*, May 7, 2012.

79. I use the term *foreign oil companies* to refer to international oil companies and national oil companies that invest overseas.

80. See Brooks, *Producing Security*, 59, for a general version of this argument.

81. Bolivia offered 10 percent, Mexico 25 percent. The Iranian government initially offered nothing to the company it dispossessed. Alan W. Ford, *The Anglo-Iranian Oil Dispute of 1951–1952* (Berkeley: University of California Press, 1954), 212.

82. Stanley Reed and Raphael Minder, "Repsol in $5 Billion Settlement with Argentina," *New York Times*, February 25, 2014.

83. Alexandra Ulmer, "Order for Venezuela to Pay Exxon $1.4 Billion in Damages Overturned," Reuters, March 9, 2017.

84. See Brooks, *Producing Security*, 60, for a general version of this argument.

85. Giroux, Burgherr, and Melkunaite, “Research Note,” 120.

86. Bala-Gbogbo, “Shell to Chevron.”

87. Kim Lewis, “Conflict Stifles South Sudan’s Oil Industry,” *Voice of America*, July 13, 2015.

88. Stanley Reed and Clifford Krauss, “New Sanctions to Stall Exxon’s Arctic Plans,” *New York Times*, September 12, 2014.

89. Lee finds that, for intrastate conflicts, the level of decline is conditioned by oil prices. However, the effect is persistent for international conflicts, regardless of oil price. Hoon Lee, “Does Armed Conflict Reduce Foreign Direct Investment in the Petroleum Sector?,” *Foreign Policy Analysis* 13, no. 1 (2016): 188–214.

90. This principle appears most prominently in the UN Charter, art. 2, para. 4, but has also become part of customary international law.

4. SEARCHING FOR CLASSIC OIL WARS

1. Other authors who argue that leaders are cost conscious when it comes to decisions about oil-related international aggression include Omar S. Bashir, “The Great Games Never Played: Explaining Variation in International Competition over Energy,” *Journal of Global Security Studies* 2, no. 4 (2017): 288–306; and Rosemary A. Kelanic, “The Petroleum Paradox: Oil, Coercive Vulnerability, and Great Power Behavior,” *Security Studies* 25, no. 2 (2016): 181–213.

2. Glenn Palmer et al., “The MID4 Dataset, 2002–2010: Procedures, Coding Rules and Description,” *Conflict Management and Peace Science* 32, no. 2 (2015): 222–42.

3. MIDs can last from one day to multiple years and often involve a series of militarized incidents.

4. Indra de Soysa, Erik Gartzke, and Tove Grete Lie, “Oil, Blood, and Strategy: How Petroleum Influences Interstate Conflict” (unpublished manuscript, April 15, 2011), 4. This threshold is employed in the Peace Research Institute Oslo (PRIO) Armed Conflict Dataset on intrastate violence and is recorded in the MID data set.

5. *Territorial* and *other* MIDs are not a universal sample of potential classic oil wars. The MID data set classifies conflicts over maritime areas as *policy* disputes. Consequently, excluding *policy* MIDs from my analysis could omit conflicts over offshore oil resources. This risk is small, however, since most interstate disagreements over offshore resources also entail disputes over islands, which the data set codes as *territorial.* Moreover, to ensure that my analysis did not overlook any classic oil wars, I examined all *policy* MIDs from 1912 to 2010 that resulted in more than twenty-five battle deaths. None of the one hundred cases examined in this robustness check were classic oil wars.

6. Daniel Yergin, *The Prize: The Epic Quest for Oil, Money, and Power* (New York: Touchstone, 1992), 156.

7. Alex Braithwaite, “MIDLOC: Introducing the Militarized Interstate Dispute Location Dataset,” *Journal of Peace Research* 47, no. 1 (2010): 91–98. Sources for territorial claims included Ewan W. Anderson, *International Boundaries: A Geopolitical Atlas* (New York: Routledge, 2003); Jacob Bercovitch and Judith Fetter, *Regional Guide to International Conflict and Management from 1945 to 2003*, rev. ed. (Washington DC: CQ, 2004); Peter Calvert, ed., *Border and Territorial Disputes of the World*, 4th ed. (London: John Harper, 2004); Paul K. Huth and Todd L. Allee, *The Democratic Peace and Territorial Conflict in the Twentieth Century* (Cambridge: Cambridge University Press, 2002); Victor Prescott and Clive Schofield, *The Maritime Political Boundaries of the World*, 2nd ed. (Leiden: Martinus Nijhoff, 2005); Victor Prescott and Gillian D. Triggs, *International Frontiers and Boundaries: Law, Politics, and Geography* (Leiden: Martinus Nijhoff, 2008); Uppsala Conflict Data Program, Conflict Encyclopedia, accessed June 24, 2019, http://www.ucdp.uu.se/gpdatabase/search.php; and the Issue Correlates of War data set presented in Bryan A. Frederick,

Paul R. Hensel, and Christopher Macaulay, "The Issue Correlates of War Territorial Claims Data, 1816–2001," *Journal of Peace Research* 54, no. 1 (2017): 99–108. I adopted this more inclusive approach, of taking states' claims into account, as well as conflicts' physical locations, in order to avoid omitting any oil-driven conflicts from the study. I classified MIDs as "involving hydrocarbon-endowed territories" when at least one participant was engaged in a serious effort to assert control over an area that contained known or prospective oil or natural gas resources.

8. Päivi Lujala, Jan Ketil Rød, and Nadia Thieme, "Fighting over Oil: Introducing a New Dataset," *Conflict Management and Peace Science* 24, no. 4 (2007): 239–56.

9. The Petroleum Dispute Dataset is also missing discovery dates for some fields, creating uncertainty about whether MID participants were aware of contested territories' resource endowments when some of the conflicts began.

10. Hydrocarbon beliefs did not have to be universal or accurate for a case to receive this coding. Petroleum beliefs were mistaken but widespread in the Chaco dispute (Bolivia–Paraguay), the Bakassi Peninsula dispute (Cameroon–Nigeria), and the Agacher Strip dispute (Burkina Faso–Mali).

11. While not depicted in this table, many of the dyads also prosecuted *territorial* MIDs in regions lacking hydrocarbon endowments, as well as MIDs that were not territorially motivated.

12. Most of the individual MIDs were bilateral.

13. The participant lists for the world wars consist of all hydrocarbon-endowed countries where the associated MIDs were prosecuted.

14. This figure is approximate, as the MID data set codes fatalities for some MIDs as *unknown* (-9). I investigated these cases myself to estimate fatality figures; most fell short of the twenty-five battle deaths threshold.

15. In November 2004, Japan protested a Chinese submarine sailing through the Ishigaki Strait, in Japan's territorial waters. A journalist reported that the submarine "was operating in waters near where Chinese vessels earlier this year began exploring for gas deposits." However, the exploration area was over two hundred miles away from the strait. Edward Cody, "Beijing Explains Submarine Activity," *Washington Post Foreign Service*, November 17, 2004; Peter A. Dutton, "International Law and the November 2004 'Han Incident,'" in *China's Future Nuclear Submarine Force*, ed. Andrew Erickson et al. (Annapolis, MD: Naval Institute Press, 2007), 162–63.

16. My analysis builds on the *issue approach* proposed by Paul Diehl. Paul F. Diehl, "What Are They Fighting For? The Importance of Issues in International Conflict Research," *Journal of Peace Research* 29, no. 3 (1992): 333–44.

17. Conflicts that result in fewer than twenty-five battle deaths attract limited journalistic and scholarly attention, which introduced some uncertainty into my analysis. Accordingly, other researchers might code some of my mild red herrings as oil spats or vice versa. However, since these are all minor conflicts, any coding discrepancies do not undermine the study's larger finding: that states avoid classic oil wars.

18. On descriptive typologies, see David Collier, Judy Laporte, and Jason Seawright, "Typologies: Forming Concepts and Creating Categorical Variables," in *The Oxford Handbook of Political Methodology*, ed. Janet M. Box-Steffensmeier, Henry Brady, and David Collier (Oxford: Oxford University Press, 2008), 152–73. These are also called classificatory typologies. Colin Elman, "Explanatory Typologies in Qualitative Studies of International Politics," *International Organization* 59, no. 2 (2005): 293–326.

19. "Cyprus Says Turkish Vessel Encroaching on Its Offshore Gas Areas," Reuters, October 20, 2014; "Greek Cyprus Warns Turkey to Stop 'Bullying' over Gas," *Hurriyet Daily News*, January 1, 2015; Daphne Tsagari, "Turkish 'Barbaros' Vessel Enters Cyprus EEZ Zone without Permission," *Greek Reporter*, October 20, 2014.

20. The three exceptions are MIDs 1172 (Argentina–Uruguay), 3825 (Germany–Romania), and 3185 (Turkey–United Kingdom).

21. Memorial of Tunisia for the Case concerning the Continental Shelf (Tunisia, Libyan Arab Jamahiriya), May 27, 1980; Thomas Lippman, "American Oil Rig Sparks Libya-Tunisia Quarrel," *Washington Post*, May 31, 1977.

22. Emily Meierding, "Do Countries Fight over Oil?," in *The Palgrave Handbook of the International Political Economy of Energy*, ed. Thijs Van de Graaf et al. (London: Palgrave Macmillan, 2016), 444–49.

23. Zou Keyuan, "Joint Development in the South China Sea: A New Approach," *International Journal of Marine and Coastal Law* 21, no. 1 (2006): 88–89.

24. "Border Movements on Tuesday: Venezuelan Craft Strayed into Guyana's Airspace, Shots Fired from Ankoko Island," *Stabroek News*, October 8, 1999; Patrick Denny, "Govt Alerts Sir Alister on Caracas Troop Ploys: Considers Them a Serious Matter," *Stabroek News*, October 9, 1999.

25. The fatalities occurred along the Evros River, which forms a portion of the states' land boundary. "Three Killed in Border Clash," Associated Press, December 19, 1986.

26. Daniel J. Dzurek, "Eritrea-Yemen Dispute over Hanish Islands," *IBRU Boundary and Security Bulletin*, Spring 1996, 72. Oil's contribution to this conflict is questionable.

27. After Germany occupied Romania in 1940 (MID 3825), it went on to attack the Soviet Union. However, the enlargement of a world war is not the dynamic that commentators envisage when they predict that oil spats will escalate.

28. Again, MID 3825 is somewhat anomalous, as Romania changed sides in the final year of World War II to fight with the Allies against Germany from August 1944 to May 1945.

29. The four exceptions are MIDs 1427 (Ethiopia–Somalia), 3825 (Germany–Romania), 4546 (Iran–Iraq), and 3185 (Turkey–United Kingdom).

30. Four exceptions are MIDs 4317 (Azerbaijan–Iran), 4121 (Eritrea–Yemen), 3610 (Indonesia–Vietnam), and 3014 (Libya–Tunisia). However, while the states that participated in these conflicts had not prosecuted prior MIDs, they had unresolved boundary disagreements at the time of their oil spats. In the fifth and final exception, MID 3825, Germany and Romania had previously confronted each other in World War I, in a campaign that was likely driven by Germany oil ambitions (although see note 59).

31. Krista Wiegand, *Enduring Territorial Disputes: Strategies of Bargaining, Coercive Diplomacy, and Statecraft* (Athens: University of Georgia Press, 2011).

32. Germany and Romania were neighbors by the time of their oil spat, because of Hitler's incorporation of Austria and Czechoslovakia, and alliance with Hungary.

33. Paul D. Senese, "Territory, Contiguity, and International Conflict: Assessing a New Joint Explanation," *American Journal of Political Science* 49, no. 4 (2005): 769–79.

34. Put otherwise, oil was neither a necessary nor a sufficient cause of conflict.

35. This number is also approximate, because of the –9 category referenced earlier.

36. Francesco Caselli, Massimo Morelli, and Dominic Rohner, "The Geography of Interstate Resource Wars," *Quarterly Journal of Economics* 130, no. 1 (2015): 267–315; Arthur H. Westing, "Appendix 2. Wars and Skirmishes Involving Natural Resources: A Selection from the Twentieth Century," in *Global Resources and International Conflict: Environmental Factors in Strategic Policy and Action*, ed. Arthur H. Westing (Oxford: Oxford University Press, 1986), 204–9. Another case that appears on Caselli, Morelli, and Rohner's list is Peru and Ecuador's repeated confrontations over the Cordillera del Cóndor. However, this classification mistakenly conflates two locations: the Maynas/Oriente region that the states contested before settling their international boundary in the Rio Protocol (1942), and a small boundary zone in the Cordillera del Cóndor where clashes occurred after the protocol. Rumors about oil in the Maynas/Oriente did not emerge until

after the Rio Protocol resolved the regional dispute. As for the Cordillera area, it contains no petroleum resources. Ecuador did eventually revive its more expansive territorial claim to the Maynas/Oriente. However, leaders did not seriously pursue it or engage in any militarized aggression in that area; instead, they limited their activities to the Cordillera. Ecuador's primary goal, in the Cordillera confrontations, was to gain control over the headwaters of the Cenepa River in order to obtain an outlet to the Amazon River via the Río Marañón. David Scott Palmer, "Peru-Ecuador Border Conflict: Missed Opportunities, Misplaced Nationalism, and Multilateral Peacekeeping," *Latin American Politics and Society* 39, no. 3 (1997): 109–48.

37. These issues also dominated aggressors' decision making in the mild red herrings.

38. Emily Meierding, "When Is an 'Oil War' Not about Oil? The Bakassi Dispute" (paper presented at the International Studies Association Annual Convention, New Orleans, LA, February 17–20, 2010).

39. The Malians were also eager to contain a perceived threat from Burkina Faso's revolutionary Sankara regime. Pierre Englebert, *Burkina Faso: Unsteady Statehood in West Africa* (Boulder, CO: Westview, 1996), 151, 154; Alain Maharaux, "La Haute-Volta devient Burkina Faso: Un territoire qui se crée, se défait et s'affirme au rythme des enjeux," in *Le territoire, lien ou frontière? Identités, conflits ethniques, enjeux et recompositions territoriales*, ed. Joël Bonnemaison, Luc Cambrézy, and Laurence Quinty Bourgeois (Paris: Orstrom, 1997), 8.

40. Ziv Rubinovitz and Elai Rettig, "Crude Peace: The Role of Oil Trade in the Israeli–Egyptian Peace Negotiations," *International Studies Quarterly* 62, no. 2 (2018): 371–82.

41. Chaim Herzog, *The Arab–Israeli Wars* (1982; New York: Vintage Books, 2005), 195; Ritchie Ovendale, *The Origins of the Arab–Israeli Wars*, 4th ed. (New York: Routledge, 2015), 232. One of the most widely cited works on the Six-Day War, Michael B. Oren, *Six Days of War: June 1967 and the Making of the Modern Middle East* (New York: Presidio, 2002), does not mention the Sinai's oil fields by name.

42. E. G. H. Joffe, "Libya and Chad," *Review of African Political Economy* 8, no. 21 (1981): 84–102; Oye Ogunbadejo, "Qaddafi's North African Design," *International Security* 8, no. 1 (1983): 154–78.

43. David D. Laitin and Said S. Samatar, *Somalia: Nation in Search of a State* (Boulder, CO: Westview, 1987), 129–36, 141–42; Robert Mallett, *Mussolini in Ethiopia, 1919–1935: The Origins of Fascist Italy's African War* (New York: Cambridge University Press, 2015); Gebru Tareke, *The Ethiopian Revolution: War in the Horn of Africa* (New Haven, CT: Yale University Press, 2009), 182–84.

44. Marwyn S. Samuels, *Contest for the South China Sea* (New York: Methuen, 1982), 111–12.

45. On Chinese concerns about the Soviet Union, see Chi-Kin Lo, *China's Policy towards Territorial Disputes: The Case of the South China Sea Islands* (London: Routledge, 1989), 70–73. On the confrontation more generally, see M. Taylor Fravel, *Strong Borders, Secure Nation: Cooperation and Conflict in China's Territorial Disputes* (Princeton, NJ: Princeton University Press, 2008), 280–83; and Lu Ning, *Flashpoint Spratlys!* (Singapore: Dolphin, 1995), 77–84.

46. Samuels, *Contest*, 92, 98–99, 112.

47. Selig S. Harrison, *China, Asia, and Oil? Conflict Ahead?* (New York: Columbia University Press, 1977), 193–94.

48. Ning, *Flashpoint Spratlys!*, 97–98. China was the last of the South China Sea's claimant states to do so.

49. Fravel, *Strong Borders*, 288–96; Ning, *Flashpoint Spratlys!*, 90–93. Both authors suggest that hydrocarbon resources were one incentive for China's expanded presence in the region.

50. Ogunbadejo, "Qaddafi's North African Design," 158, 175.

51. Burkina Faso described the Agacher Strip's manganese deposits in its memorial to the International Court of Justice, which was evaluating the case. Neither state mentioned oil or natural gas deposits. Mémoire de Burkina Faso, Case concerning the Frontier Dispute (Burkina Faso/Republic of Mali), October 3, 1985, 37.

52. Ralf Emmers, *Resource Management and Contested Territories in East Asia* (London: Palgrave Macmillan, 2013), 52–54.

53. Pascal James Imperato, *Mali: A Search for Direction* (Boulder, CO: Westview, 1989), 138; Meierding, "Bakassi Dispute."

54. Estonia possesses oil shale.

55. All of the historical oil campaigns targeted oil, not natural gas.

56. For example, see David A. Deese, "Oil, War, and Grand Strategy," *Orbis* 25, no. 4 (1981): 525–55; Robert Goralski and Russell W. Freeburg, *Oil and War: How the Deadly Struggle for Fuel in WWII Meant Victory or Defeat* (New York: William Morrow, 1987); and Yergin, *Prize*.

57. To identify these cases, I deviated from my focus on conflict onset, as these campaigns occurred in the midst of ongoing wars (MIDs 157, 257, and 258). I based that methodological choice on the campaigns' significance for conflict studies, generally, and for classic oil wars research, specifically. Two midwar attacks that I did not code as oil campaigns are Great Britain's assault against the Rashid Ali regime in Iraq (May 1941) and the Anglo–Soviet invasion of Iran (1941–1942). These campaigns occurred in oil-endowed territories but did not target petroleum resources. Instead, Allied aggressors aimed to discourage closer ties between the targeted regimes and Nazi Germany. In the latter incident, Anglo–Soviet intervention also secured supply routes for US lend–lease assistance to the Soviet Union. As I discuss in Chapter 7, I also do not code Germany's invasion of Romania in World War II as an oil campaign, because it was insufficiently deadly.

58. General Erich Ludendorff, the leader of Germany's war effort, claimed that oil was not a core motive for the World War I counteroffensive. As he put it, "The subjugation of Romania was a military necessity for us, and we took economic advantage of the conquest *en passant*. We should never under any circumstances have attacked Romania to seize her economic resources." Quoted in Ian O. Lesser, *Resources and Strategy* (New York: St. Martin's, 1989), 42. Given Ludendorff's position, the accuracy of his claim is suspect, so I label this an oil campaign. However, if Ludendorff was sincere, his statement provides further evidence of states' limited willingness to fight for oil.

59. Deese, "Oil, War, and Grand Strategy," 531; Kelanic, "Petroleum Paradox," 197–202; Lesser, *Resources and Strategy*, 42–45; Timothy C. Winegard, *The First World Oil War* (Toronto: University of Toronto Press, 2016). British officials aimed to control Mosul's prospective petroleum resources to improve their energy security after the war. Victor H. Rothwell, "Mesopotamia in British War Aims: 1914–1918," *Historical Journal* 13, no. 2 (1970): 286–91.

60. Jan Selby, "Oil and Water: The Contrasting Anatomies of Resource Conflicts," *Government and Opposition* 40, no. 2 (2005): 202.

61. Oil gambits are a type of "gamble for resurrection." George W. Downs and David M. Rocke, "Conflict, Agency, and Gambling for Resurrection: The Principal-Agent Problem Goes to War," *American Journal of Political Science* 38, no. 2 (1994): 362–80.

5. RED HERRINGS

1. Julian Duguid, *Green Hell: Adventures in the Mysterious Jungles of Eastern Bolivia* (New York: Century, 1931); J. Valerie Fifer, *Bolivia: Land, Location, and Politics since 1825* (London: Cambridge University Press, 1972), 162; Harris Gaylord Warren, *Paraguay: An Informal History* (Norman: University of Oklahoma Press, 1949), 300.

2. Fifer, *Bolivia*, 181–83; Gordon Ireland, *Boundaries, Possessions, and Conflicts in South America* (Cambridge, MA: Harvard University Press, 1938), 94; David H. Zook Jr., *The Conduct of the Chaco War* (New Haven, CT: Bookman, 1960), 25.

3. These included the Quijarro–Decoud Treaty (1879), Tamayo–Aceval Treaty (1887), Ichazo–Benítez Treaty (1894), and Pinilla–Soler Protocol (1907).

4. Fifer, *Bolivia*, 196n1.

5. Bruce W. Farcau, *The Chaco War: Bolivia and Paraguay, 1932–1935* (Westport, CT: Praeger, 1996), 7; League of Nations, *Dispute between Bolivia and Paraguay, Appeal of the Bolivian Government under Article 15 of the Covenant*, Extracts from the Records of the Fifteenth Ordinary Session of the Assembly, League of Nations Official Journal, Special Supplement No. 124 (Geneva, 1934), 157; Harris Gaylord Warren, *Rebirth of the Paraguayan Republic: The First Colorado Era, 1878–1904* (Pittsburg: University of Pittsburg Press, 1985), 170–72; Zook, *Conduct*, 27.

6. Bridget María Chesterton, *The Grandchildren of Solano López: Frontier and Nation in Paraguay, 1904–1936* (Albuquerque: University of New Mexico Press, 2013), 6, 97–101; Farcau, *Chaco War*, 11; Fifer, *Bolivia*, 184, 208; Ronald Stuart Kain, "Behind the Chaco War," *Current History* 42, no. 5 (1935): 469; Leslie B. Rout Jr., *Politics of the Chaco Peace Conference, 1935–1939* (Austin: University of Texas Press, 1970), 13; Zook, *Conduct*, 37.

7. Stephen C. Cote, *Oil and Nation: A History of Bolivia's Petroleum Sector* (Morgantown: West Virginia University Press, 2016), 5, 16–18, 49; Stephen C. Cote, "Bolivian Oil Nationalism and the Chaco War," in *The Chaco War: Environment, Ethnicity, and Nation*, ed. Bridget María Chesterton (London: Bloomsbury Academic, 2016), 158–63; Herbert S. Klein, "American Oil Companies in Latin America: The Bolivian Experience," *Inter-American Economic Affairs* 18, no. 2 (1964): 47–55; Herbert S. Klein and José Alejandro Peres-Cajías, "Bolivian Oil and Natural Gas under State and Private Control, 1920–2010," *Bolivian Studies Journal/Revista de Estudios Bolivianos* 20 (2014): 147.

8. Stephen Cote, "A War for Oil in the Chaco, 1932–1935," *Environmental History* 18, no. 4 (2013): 4; Cote, "Bolivian Oil Nationalism," 161; Cote, *Oil and Nation*, 54; Klein, "American Oil Companies," 55n27.

9. Imprenta Militar, *Paraguay-Bolivia: Aspectos de la guerra del Chaco* (Asunción: Publicaciones del Servicio de Información y Propaganda del Ministerio de la Defensa Nacional, 1934), 23; J. W. Lindsay, "The War over the Chaco: A Personal Account," *International Affairs* 14, no. 2 (1934): 233–34; Kain, "Behind the Chaco War," 470–71. Bolivia began construction of a Cochabamba–Santa Cruz railway line in 1928 but abandoned it because of the cost.

10. La Paz was reluctant to improve connections between its eastern provinces and Argentina before a strong linkage had been established between its own east and west. Buenos Aires wanted to expand its own oil industry and had little interest in bringing competing Bolivian products into the international marketplace. Argentina was especially hostile because the oil was being produced by Standard; officials had clashed with the company earlier in the 1920s. J. Valerie Fifer, "Bolivia's Pioneer Fringe," *Geographical Review* 57 (1967): 22–23; Fifer, *Bolivia*, 191–92; Rout, *Politics*, 56–58.

11. Cote, *Oil and Nation*, 79; Kain, "Behind the Chaco War," 473; Ronald Stuart Kain, "Bolivia's Claustrophobia," *Foreign Affairs* 16, no. 4 (July 1938): 709; Rout, *Politics*, 47; Alfredo M. Seiferheld, *Economía y petróleo durante la guerra del Chaco: Apuntes para una historia económica del conflicto Paraguayo-Boliviano* (Asunción: Instituto Paraguayo de Estudios Geopolíticos e Internacionales, 1983), 473–74.

12. Farcau, *Chaco War*, 11, 18–19; Fifer, *Bolivia*, 194–96, 207; Zook, *Conduct*, 37, 62–64.

13. Chesterton, *Grandchildren of Solano López*, 4–5, 32; Zook, *Conduct*, 43–44.

14. Paul R. Hensel, "Evolution in Domestic Politics and the Development of Rivalry: The Bolivia-Paraguay Case," in *Evolutionary Interpretations of World Politics*, ed. William R. Thompson (New York: Routledge, 2001), 197.

15. Zook, *Conduct*, 50, 5. See also Chesterton, *Grandchildren of Solano López*, 110–12.

16. Bolivia calls the lake Chuquisaca; Paraguay calls it Pitiantuta.

17. Quoted in Farcau, *Chaco War*, 37. See also Ireland, *Boundaries*, 75; and Zook, *Conduct*, 69–73.

18. The war was not formally declared until 1933.

19. Paraguayan and Argentine newspapers started accusing Standard of provoking the war "almost as soon as the fighting started." Cote, *Oil and Nation*, 70, quoting Seiferheld, *Economía*. See also Klein, "American Oil Companies," 57; Klein and Peres-Cajías, "Bolivian Oil," 147; Warren, *Paraguay*, 297; and Kevin A. Young, *Blood of the Earth: Resource Nationalism, Revolution, and Empire in Bolivia* (Austin: University of Texas Press, 2017), 21.

20. Cote, "Bolivian Oil Nationalism," 164; Seiferheld, *Economía*, 459–62. Tristán Marof (the pen name of Gustavo Adolpho Navarro) was one of the most prominent intellectuals blaming the war on multinational oil companies.

21. These include Hans J. Morgenthau, *Politics among Nations: The Struggle for Power and Peace*, 4th ed. (1948; New York: Alfred A. Knopf, 1967), 46.

22. Senator Long (LA), "The Chaco," *Congressional Record*, June 8, 1934, 10808, 10811–12.

23. League of Nations, *Dispute*, 155–56.

24. Michael L. Gillette, "Huey Long and the Chaco War," *Louisiana History: The Journal of the Louisiana Historical Association* 11, no. 4 (1970): 302.

25. Enrique Finot, *The Chaco War and the United States* (New York: L. and S., 1934), 11–13.

26. Ireland, *Boundaries*, 94; Kain, "Behind the Chaco War," 469. Later analyses of the conflict concurred. Historian Bryce Wood wrote that "the Chaco itself was not known to possess important mineral resources," while Herbert S. Klein asserted that "absolutely no oil lands were at stake in the Chaco territory dispute," and David H. Zook stated that "oil was specifically insignificant in the origins of the Chaco war." Bryce Wood, *The United States and Latin American Wars, 1932–1942* (New York: Columbia University Press, 1966), 20; Herbert S. Klein, *Parties and Political Change in Bolivia, 1880–1952* (London: Cambridge University Press, 1969), 153; Zook, *Conduct*, 72.

27. Senator Logan (KY), *Congressional Record*, January 28, 1935, 1116.

28. Senator Long (LA), *Congressional Record*, January 29, 1935, 1122.

29. Cote, *Oil and Nation*, 78–81; Cote, "Bolivian Oil Nationalism," 157, 163; Klein, "American Oil Companies," 57, 63; Klein and Peres-Cajías, "Bolivian Oil," 148n6; Warren, *Paraguay*, 194.

30. Finot, *Chaco War*.

31. Imprenta Militar, *Paraguay-Bolivia*, 22.

32. For example, William L. Schurz, "The Chaco Dispute between Bolivia and Paraguay," *Foreign Affairs* 7, no. 4 (1929): 650–55.

33. Gillette, "Huey Long," 299.

34. I include Bolivia and Paraguay's Chaco confrontations on the list of militarized interstate disputes involving hydrocarbon-endowed territories based on these popular assumptions. Although incorrect, and not shared by high government officials, the beliefs were widespread.

35. Senator Long (LA), "The Chaco," *Congressional Record*, June 8, 1934, 10808–12; Senator Long (LA), *Congressional Record*, January 7, 1935, 563, 567.

36. Gillette, "Huey Long."

37. Senator Long (LA), *Congressional Record*, June 7, 1934, 10811–12.

38. This accusation appeared in the state's submission to the League of Nations. League of Nations, *Dispute*, 111.

39. Cote, "War for Oil," 2.

40. Cote, *Oil and Nation*, 82–83.

41. Manuel E. Contreras, "The Bolivian Tin Mining Industry in the First Half of the Twentieth Century" (University of London, Institute of Latin American Studies Research Papers, No. 32, 1993), 3, 8, 11–15.

42. Cote, "War for Oil," 10–11. They also unilaterally altered Standard's contract to increase royalty payments. Cote, "Bolivian Oil Nationalism," 161–63.

43. Eduardo Arze Quiroga, ed., *Documentos para una historia de la guerra del Chaco: Archivos de Daniel Salamanca* (La Paz: Editorial Don Bosco, 1951), 1:157.

44. Kain, "Bolivia's Catastrophe," 704; Kain, "Behind the Chaco War," 468.

45. Translated from Seiferheld, *Economía*, 475.

46. Senator Long (LA), *Congressional Record*, January 7, 1935, 566.

47. Jeff D. Colgan, "Fueling the Fire: Pathways from Oil to War," *International Security* 38, no. 2 (2013): 147–80.

48. Fifer, *Bolivia*, 200; Ireland, *Boundaries*, 70; Kain, "Bolivia's Catastrophe," 709; Rout, *Politics*, 46.

49. Farcau, *Chaco War*, 8, 14; Rout, *Politics*, 22n35, 50–51n25; Warren, *Rebirth*, 154.

50. Rout, *Politics*, 51.

51. "Letter from the Chief of the Bolivian General Staff to Salamanca, August 1932," document 36 in Quiroga, *Documentos*, 333–41.

52. Schurz, "Chaco Dispute," 653; Farcau, *Chaco War*, 8–9; Fifer, *Bolivia*, 28–30, 260.

53. Bolivia and Chile signed a truce in 1884, which allowed Chile to occupy Bolivia's coastal territories, and a settlement in 1904, which formally transferred the province to Chile but gave Bolivia "guaranteed commercial transit rights." Fifer, *Bolivia*, 66.

54. The Bolivian vice president called the Tacna–Arica decision a "final blow" to national prestige. Quoted in Rout, *Politics*, 27.

55. The phrase is from Paraguay's submission to the League of Nations, which presented it mockingly and asserted that Bolivia's isolation was a "myth." League of Nations, *Dispute*, 155. On Bolivia's efforts to acquire a sea outlet via the Amazon, see Fifer, *Bolivia*, 104–36.

56. Rout, *Politics*, 22, 24.

57. Cote, "War for Oil," 4; Farcau, *Chaco War*, 9; Hensel, "Evolution," 194.

58. Farcau, *Chaco War*, 24. See also Wood, *United States*, 24.

59. Rout, *Politics*, 24.

60. Fifer, *Bolivia*, 196; Rout, *Politics*, 13, 17; Zook, *Conduct*, 30.

61. Chesterton, *Grandchildren of Solano López*, 7–8, 33–34; Paul H. Lewis, *Political Parties and Generations in Paraguay's Liberal Era, 1869–1940* (Chapel Hill: University of North Carolina Press, 1993), 152–53; Rout, *Politics*, 16–18.

62. Zook, *Conduct*, 38. See also Chesterton, *Grandchildren of Solano López*, 38–55; Hensel, "Evolution," 195–96.

63. Chesterton, *Grandchildren of Solano López*, 57, 110–11; Bridget María Chesterton, "Introduction: An Overview of the Chaco," in Chesterton, *Chaco War*, 5–7; Farcau, *Chaco War*, 9, 26; Fifer, *Bolivia*, 209; Rout, *Politics*, 9, 18; Schurz, "Chaco Dispute," 654; Zook, *Conduct*, 38, 46–47, 50, 54.

64. Herbert S. Klein makes this domestic argument most strongly: for example, Klein, *Parties*, 133–35, 139–45, 152; and Herbert S. Klein, "The Crisis of Legitimacy and the Origins of Social Revolution: The Bolivian Experience," *Journal of Inter-American Studies* 10, no. 1 (1968): 102–16. See also Zook, *Conduct*, 58.

65. Farcau, *Chaco War*, 13. Paraguay's submission to the League of Nations included a quotation from Salamanca: "It [Paraguay] is the only country we can attack with the certainty of victory." League of Nations, *Dispute*, 151.

66. Farcau, *Chaco War*, 11–13; Klein, *Parties*, 130; Zook, *Conduct*, 33.

67. Initially, many Bolivians were enthusiastic about the war. Herbert S. Klein, *A Concise History of Bolivia* (Cambridge: Cambridge University Press, 2003), 178. However, much of the population turned against Salamanca by October 1932, after significant Bolivian military defeats. Cote, *Oil and Nation*, 75–76.

68. Chesterton, *Grandchildren of Solano López*, 6; Fifer, *Bolivia*, 209, 219; Ireland, *Boundaries*, 75.

69. Herbert S. Klein, *Orígenes de la revolución nacional boliviana: La crisis de la generación del Chaco* (La Paz: Juventad, 1968), 220. Translated from the original: "Esta mitologia de la guerra por el petróleo vino a ser universalmente aceptada por todos."

70. Many authors include the conflict on their lists of classic oil wars. See, for example, Francesco Caselli, Massimo Morelli, and Dominic Rohner, "The Geography of Interstate Resource Wars," *Quarterly Journal of Economics* 130, no. 1 (2015): 267; Colgan, "Fueling the Fire," 154; and Daniel Moran and James A. Russell, "Introduction: The Militarization of Energy Security," in *Energy Security and Global Politics: The Militarization of Resource Management*, ed. Daniel Moran and James A. Russell (New York: Routledge, 2009), 17.

71. Andrew T. Price-Smith, *Oil, Illiberalism, and War: An Analysis of Energy and US Foreign Policy* (Boston: MIT Press, 2015), 82.

72. Sources are, in order, "Iraq, Iran." Foreign Broadcast Information Service (FBIS), Daily Report, Middle East and Africa, FBIS-MEA-80-178, September 11, 1980, i; Efraim Karsh, "Military Power and Foreign Policy Goals: The Iran–Iraq War Revisited," *International Affairs* 24, no. 1 (1987/1988): 94; "Text of President's Speech to National Assembly 17 Sep," Baghdad Domestic Service in Arabic (1607 GMT, September 17, 1980), FBIS-MEA-80-183, September 18, 1980, E1–7; and "Revolution Command Council Statement on Military Actions," Baghdad Domestic Service in Arabic (1143 GMT, September 22, 1980), FBIS-MEA-80-186, September 23, 1980, E8–9.

73. Sources are, in order, "Baghdad Radio Broadcasts in Persian to Iran 28–30 Sep," Baghdad in Persian to Iran (0730-1030 GMT, September 28, 1980), FBIS-MEA-80-192, October 1, 1980, E1"; "Statement on Principles," Baghdad Voice of the Masses in Arabic (1245 GMT, September 26, 1980), FBIS-MEA-80-191, September 30, 1980, E6; and "Foreign Ministry Official Addresses UN Security Council," Baghdad INA in Arabic (0825 GMT, September 27, 1980), FBIS-MEA-80-191, September 30, 1980, E8–17.

74. "Defense Minister Reports on Iraqi–Iranian Dispute," Baghdad INA in Arabic (1310 GMT, September 25, 1980), FBIS-MEA-80-189, September 26, 1980, E10–17.

75. Robert Litwak, *Security in the Persian Gulf 2: Sources of Inter-state Conflict*, International Institute for Strategic Studies (Montclair, NJ: Allanheld, Osmun, 1981), 13; Ibrahim Anvari Tehrani, "Iraqi Attitudes and Interpretation of the 1975 Agreement," in *The Iran-Iraq War: The Politics of Aggression*, ed. Farhang Rajaee (Gainesville: University Press of Florida, 1993), 12.

76. See, for example, Iranian prime minister Mohammed Ali Rajai's statement before the United Nations Security Council. United Nations, Security Council Official Records, 2251st meeting (New York), October 17, 1980.

77. Quoted in *Keesing's Contemporary Archives*, vol. 27 (London: Longman Group, 1981), 31015.

78. "Revolution Command Council Statement on Military Actions," FBIS-MEA-80-186.

79. "Le Monde Reports on Tariq ʿAziz News Conference," Paris Le Monde in French (September 27, 1980), FBIS-MEA-80-191, September 30, 1980, E4–5; "Tariq ʿAziz Holds

Press Conference in Amman 26 Sep," Amman Domestic Television Service in Arabic (1823 GMT, September 26, 1980), FBIS-MEA-80-190, September 29, 1980, E18–20.

80. "Text of Saddam Husayn's 10 Nov Press Conference," Baghdad INA in Arabic (1130 GMT, November 11, 1980), FBIS-MEA-80-220, November 12, 1980, E2–19.

81. "Saddam Husayn's 10 Nov Press Conference," FBIS-MEA-80-220. On using Khuzestan as a bargaining chip, see also Will D. Swearingen, "Geopolitical Origins of the Iran-Iraq War," *Geographical Review* 78, no. 4 (1988): 415.

82. Quoted in Claudia Wright, "Implications of the Iraq-Iran War," *Foreign Affairs* 59, no. 2 (1980): 288.

83. Saddam quoted in Jasim M. Abdulghani, *Iraq and Iran: The Years of Crisis* (Baltimore: Johns Hopkins University Press, 1984), 155.

84. "Saddam and His Advisers Discussing Iraq's Decision to Go to War with Iran," September 16, 1980, SH-SHTP-A-000-835, Saddam Hussein Regime Collection, Conflict Records Research Center, National Defense University, Washington, DC. This document, like others cited in this chapter, is from the Iraqi regime's archives and was acquired by US forces during the 2003 invasion of Iraq.

85. "Saddam and His Inner Circle Discussing Relations with Various Arab States, Russia, China, and the United States," undated (circa November 4–20, 1979), SH-SHTP-D-000-559, Saddam Hussein Regime Collection; "The Arabistan (Arabs in Southern Iran) in Al-Ahwaz Area Calling for Independence," March 1979–May 2000, SH-GMID-D-000-620, Saddam Hussein Regime Collection.

86. Quoted in Tareq Y. Ismael, *Iraq and Iran: Roots of Conflict* (Syracuse, NY: Syracuse University Press, 1982), 91.

87. Marion Farouk-Sluglett and Peter Sluglett, *Iraq since 1958: From Revolution to Dictatorship* (New York: I. B. Tauris, 1990), 256.

88. Phebe Marr, *The Modern History of Iraq*, 2nd ed. (Boulder, CO: Westview, 2003), 159–61.

89. "Defense Minister Reports," FBIS-MEA-80-189.

90. Edgar O'Ballance, *The Gulf War* (London: Brassey's Defence, 1988), 55.

91. The "usurped" terminology was Saddam's. For example, "Iraq Ends 1975 Border Pact with Iran as Frontier Clashes Continue," *New York Times*, September 18, 1980.

92. For a version of this argument, see F. Gregory Gause III, "Iraq's Decisions to Go to War, 1980 and 1990," *Middle East Journal* 56, no. 1 (2002): 47–70.

93. On the ascriptive divide, see Majid Khadduri, *The Gulf War: The Origins and Implications of the Iran–Iraq Conflict* (New York: Oxford University Press, 1988).

94. Shaul Bakhash, "The Troubled Relationship: Iran and Iraq, 1930–80," in *Iran, Iraq, and the Legacies of War*, ed. Lawrence G. Potter and Gary G. Sick (New York: Palgrave, 2004), 15–16; Daniel Pipes, "A Border Adrift: Origins of the Conflict," in *The Iran-Iraq War: New Weapons, Old Conflicts*, ed. Shirin Tahir-Kheli and Shaheen Ayubi (New York: Praeger, 1983), 18; Tehrani, "Iraqi Attitudes," 12. Qasim also threatened, in June 1961, to invade Kuwait, and he was the first Arab leader to use the term *Arabian Gulf* rather than Persian Gulf. Bakhash, "Troubled Relationship," 16.

95. Bakhash, "Troubled Relationship," 16–19; Karsh, "Military Power," 84; Efraim Karsh, "From Ideological Zeal to Geopolitical Realism: The Islamic Republic and the Gulf," in *The Iran–Iraq War: Impact and Implications*, ed. Efraim Karsh (Houndsmills, UK: Macmillan, 1989), 26; O'Ballance, *Gulf War*, 10; Pierre Razoux, *The Iran-Iraq War* (Cambridge, MA: Harvard University Press, 2015), 52–53.

96. Bakhash, "Troubled Relationship," 19; Litwak, *Security*, 5; Karsh, "Ideological Zeal," 28; Hussein Sirriyeh, "Development of the Iraqi-Iranian Dispute, 1847–1975," *Journal of Contemporary History* 20, no. 3 (1985): 487.

97. Dilip Hiro, *The Longest War* (New York: Routledge, 1991), 16–17.

98. Williamson Murray and Kevin M. Woods, *The Iran–Iraq War: A Military and Strategic History* (Cambridge: Cambridge University Press, 2014); Razoux, *Iran-Iraq War*, 56.

99. Quoted in Nita Renfrew, "Who Started the War?," *Foreign Policy* 66 (April 1987): 100.

100. Razoux, *Iran-Iraq War*, 57. See also "Saddam and His Advisers"; Chaim Herzog, "A Military-Strategic Overview," in Karsh, *Iran–Iraq War*, 257; Murray and Woods, *Iran–Iraq War*, 22; and Swearingen, "Geopolitical Origins," 408.

101. Bakhash, "Troubled Relationship," 13.

102. Edmund Ghareeb, "The Roots of the Crisis: Iraq and Iran," in *The Persian Gulf War: Lessons for Strategy, Law, and Diplomacy*, ed. Christopher C. Joyner (Westport, CT: Greenwood, 1990), 26; *Iran–Iraq Boundary*, International Boundary Study No. 164 (Washington, DC: Office of the Geographer, Bureau of Intelligence and Research, 1978), 3–4; Richard Schofield, "Position, Function, and Symbol: The Shatt al-ʿArab Dispute in Perspective," in Potter and Sick, *Legacies of War*, 36–39.

103. Bakhash, "Troubled Relationship," 13–16; Pipes, "Border Adrift," 18–19; Schofield, "Position, Function, and Symbol," 52.

104. Bakhash, "Troubled Relationship," 18, 20; Karsh, "Military Power," 86; Litwak, *Security*, 4; Pipes, "Border Adrift," 19–20; Sirriyeh, "Development," 485–87.

105. Murray and Woods, *Iran–Iraq War*, 22; Swearingen, "Geopolitical Origins," 415. The Iraqis were not attempting to secure oil transportation. Although some oil equipment was shipped via the Shatt al-ʿArab to Basra, most crude oil and petroleum products from Iraq's southern fields were transported to the state's Gulf export terminals by pipeline, because of the waterway's limited draft and the increasing size of crude carriers.

106. Abdulghani, *Iraq and Iran*, 201; "Foreign Minister Makes Statement on Dispute with Iran," Baghdad INA in Arabic (2014 GMT, September 10, 1980)," FBIS-MEA-80-178, September 11, 1980, E2–3; "Saddam and His Advisers," SH-SHTP-A-000-835.

107. Anthony H. Cordesman and Abraham R. Wagner, *The Lessons of Modern War*, vol. 2, *The Iran-Iraq War* (Boulder, CO: Westview, 1990), 21.

108. Tehrani, "Iraqi Attitudes," 16. In this process, they were guided by the Treaty on International Borders and Good Neighborly Relations, which the states signed on June 13, 1975, as a follow-up to the Algiers Agreement. According to this treaty, the boundary would be redemarcated on the basis of the Constantinople Protocol (1913), the minutes of the Turkish–Persian Border Demarcation Commission (1914), the Teheran Protocol (March 15, 1975), and the minutes of the Iranian and Iraqi foreign ministers' talks held in spring 1975.

109. Quoted in "Foreign Minister Makes Statement," FBIS-MEA-80-178. The foreign minister made a similar statement before the United Nations on October 17. For additional, similar Iraqi statements, see "'Izzat Ibrahim Holds Press Conference in Rome 15 Sep," Baghdad INA in Arabic," (1515 GMT, September 15, 1980), FBIS-MEA-80-181, September 16, 1980, E2–3; "President's Speech to National Assembly," FBIS-MEA-80-183; and "Defense Minister Reports," FBIS-MEA-80-189.

110. On September 17, 1980, Banisadr acknowledged that Iran had failed to return the border areas as instructed by the Algiers Agreement. His attitude toward the accord was dismissive; he "asked rhetorically, 'Who signed that agreement. Even the Shah's regime did not apply it.'" Abdulghani, *Iraq and Iran*, 203.

111. Cordesman and Wagner, *Lessons of Modern War*, 30; Ghareeb, "Roots of the Crisis," 34; Hiro, *Longest War*, 39; O'Ballance, *Gulf War*, 30; Kevin M. Woods, David D. Palkki, and Mark E. Stout, eds., *The Saddam Tapes: The Inner Workings of a Tyrant's Regime, 1978–2001* (Cambridge: Cambridge University Press, 2011), 132n18.

112. "Saddam and His Advisers," SH-SHTP-A-000-835, translation from Woods, Palkki, and Stout, *Saddam Tapes*, 132.

113. "Saddam and His Advisers," SH-SHTP-A-000-835. See also Murray and Woods, *Iran–Iraq War*, 48–49.

114. "Saddam and His Advisers," SH-SHTP-A-000-835; Cordesman and Wagner, *Lessons of Modern War*, 30–31; Schofield, "Position, Function, and Symbol," 54.

115. "Summaries and Intelligence Reports for the General Military Intelligence Directorate (GMID)," September 1980–May 1985, SH-GMID-D-000-332, Saddam Hussein Regime Collection; Murray and Woods, *Iran–Iraq War*, 92.

116. Quoted in Karsh, "Military Power," 87. See also Ghareeb, "Roots of the Crisis," 30.

117. Ghareeb, "Roots of the Crisis," 31–32.

118. O'Ballance, *Gulf War*, 27.

119. Karsh, "Military Power," 87; Murray and Woods, *Iran–Iraq War*, 44, 88; Renfrew, "Who Started the War?," 99.

120. "Iran and Iraq Press the War of Words," *New York Times*, April 9, 1980; Abdulghani, *Iraq and Iran*, 189; *The Iraqi–Iranian Dispute: Facts v. Allegations* (New York: Ministry of Foreign Affairs of the Republic of Iraq, 1980), 33; Razoux, *Iran-Iraq War*, 2.

121. Murray and Woods, *Iran–Iraq War*, 44.

122. Karsh, "Military Power," 88.

123. Karsh, 88; Murray and Woods, *Iran–Iraq War*, 46, 90; Kevin M. Woods et al., *Saddam's Generals: Perspectives of the Iran-Iraq War* (Alexandria, VA: Institute for Defense Analysis, 2011), 52, 88–89, 115–17.

124. Gause, "Iraq's Decisions," 63–69; Murray and Woods, *Iran–Iraq War*, 90.

125. Cordesman and Wagner, *Lessons of Modern War*, 29; Hiro, *Longest War*, 36; Razoux, *Iran-Iraq War*, 4–6; Wright, "Implications," 280.

126. Razoux, *Iran-Iraq War*, 10; Woods et al., *Saddam's Generals*, 115.

127. " Saddam Husayn's 10 Nov Press Conference," FBIS-MEA-80-220; *Iraqi–Iranian Dispute*, 11; Murray and Woods, *Iran–Iraq War*, 47–48, 91; Renfrew, "Who Started the War?," 102.

128. "Saddam and Senior Iraqi Officials Discussing the Conflict with Iran, Iraqi Targets and Plans, a Recent Attack on the Osirak Reactor, and Various Foreign Countries," October 1, 1980, SH-MISC-D-000-827, Saddam Hussein Regime Collection; Woods et al., *Saddam's Generals*, 32; Razoux, *Iran-Iraq War*, 10–11.

129. Murray and Woods, *Iran–Iraq War*, 129; Woods et al., *Saddam's Generals*, 7.

130. Razoux, *Iran-Iraq War*, 6; Woods et al., *Saddam's Generals*, 9, 52, 54.

131. Woods et al., *Saddam's Generals*, 199. See also Farzin Nadimi, "The Role of Oil in the Outcome of the Iran–Iraq War: Some Important Lessons in Historical Context," in *The Iran–Iraq War: New International Perspectives*, ed. Nigel Ashton and Bryan Gibson (London: Routledge, 2013), 80–81; Razoux, *Iran-Iraq War*, 10–11; and "Saddam and Senior Iraqi Officials."

132. Discussing the oil motive, Renfrew claims that "Iraqi officials only half admit it and then only off the record." Pipes offers only suppositions; he asserts that "Baghdad did not make control of this province one of its stated goals, but Iraqi leaders surely aspire to conquer it" partly for oil-related reasons, and "even if taking Khuzistan [*sic*] is only a remote possibility . . . the possible advantages to Iraq are so great that surely they must have entered into Baghdad's calculations and added importantly to the other reasons for going to war in September 1980." None of these authors offer any evidence to support their classic oil war claims. Renfrew, "Who Started the War?," 103; Pipes, "Border Adrift," 22–23.

133. Nigel Ashton and Bryan Gibson, introduction to Ashton and Gibson, *Iran–Iraq War*, 6.

134. Razoux, *Iran-Iraq War*, 10–11.

6. OIL SPATS

1. Francesco Caselli, Massimo Morelli, and Dominic Rohner, "The Geography of Interstate Resource Wars," *Quarterly Journal of Economics* 130, no. 1 (2015): 267; William Safire, "In Defeat, Defiance," *New York Times*, April 5, 1982; Arthur H. Westing, "Appendix 2. Wars and Skirmishes Involving Natural Resources: A Selection from the Twentieth Century," in *Global Resources and International Conflict: Environmental Factors in Strategic Policy and Action*, ed. Arthur H. Westing (Oxford: Oxford University Press, 1986), 208–9.

2. For example, Peter Wilby, "The Islands of Black Gold," *New Statesman*, March 15, 2010.

3. Peter Beck, *The Falkland Islands as an International Problem* (London: Routledge, 1988), 43–44.

4. Aaron Donaghy, *The British Government and the Falkland Islands, 1974–79* (Houndsmills, UK: Palgrave Macmillan, 2014), 10; Fritz L. Hoffmann and Olga Mingo Hoffmann, *Sovereignty in Dispute: The Falklands/Malvinas, 1493–1982* (Boulder, CO: Westview, 1984), 95–99.

5. Martín Abel Gonzáles, *The Genesis of the Falklands (Malvinas) Conflict: Argentina, the UK, and the Failed Negotiations of the 1960s* (Houndsmills, UK: Palgrave Macmillan, 2013).

6. In 1966, the Colonial Office became the Commonwealth Office, which was absorbed into the Foreign Office in 1968.

7. Peter Beck, "Cooperative Confrontation in the Falkland Islands Dispute: The Anglo-Argentine Search for a Way Forward, 1968–1981," *Journal of Interamerican Studies and World Affairs* 24, no. 1 (1982): 38; Lawrence Freedman, *The Official History of the Falklands Campaign*, vol. 1, *The Origins of the Falklands War* (Milton Park, UK: Routledge, 2005), 18.

8. Michael Charlton, *The Little Platoon: Diplomacy and the Falklands Dispute* (Oxford: Basil Blackwell, 1989), 7–8.

9. A 1970 position paper expressed this view. Klaus Dodds, *Pink Ice: Britain and the South Atlantic Empire* (London: I. B. Tauris, 2002), 157. It was reiterated by subsequent British governments. For examples, see Lowell Gustafson, *The Sovereignty Dispute over the Falkland (Malvinas) Islands* (New York: Oxford University Press, 1988), 83.

10. Dodds, *Pink Ice*, 156.

11. Gustafson, *Sovereignty Dispute*, 83–87; Adolpho Silenzi de Stagni, *Las Malvinas y el petróleo* (Buenos Aires: El Cid Editor, 1982), 1:25.

12. Dodds, *Pink Ice*, 152–56; Gustafson, *Sovereignty Dispute*, 89; Silenzi de Stagni, *Las Malvinas*, 73.

13. Hugh O'Shaughnessy, "Britain Denies 'Go It Alone' Plans for Falkland's Oil," *Financial Times*, December 10, 1975.

14. Gustafson, *Sovereignty Dispute*, 87.

15. O'Shaughnessy, "Britain Denies."

16. Quoted in Charlton, *Little Platoon*, 42.

17. "Commercial Krill," *Financial Times*, July 21, 1976; Hugh O'Shaughnessy, "Decision on Falkland Oil Licenses Expected," *Financial Times*, April 3, 1975.

18. Hoffman and Hoffman, *Sovereignty in Dispute*, 123–24.

19. Silenzi de Stagni, *Las Malvinas*, 76. The article was entitled "Fear of Falklands Oil War."

20. Quoted in Freedman, *Official History*, 43. See also Donaghy, *British Government*, 78–81, 87–88, 105.

21. Donaghy, *British Government*, 84–85.

22. Freedman, *Official History*, 44–45.

23. Donaghy, *British Government*, 82–88; Freedman, *Official History*, 44–45, 54–55; Hoffman and Hoffman, *Sovereignty in Dispute*, 126–28.

24. Quoted in Donaghy, *British Government*, 89.

25. "Argentina Wants Envoy Withdrawn," *Times* (London), January 4, 1976; Donaghy, *British Government*, 89–90.

26. Oliver Franks, *Falkland Islands Review: Report of a Committee of Privy Councilors* (London: Falkland Islands Review Committee, Her Majesty's Stationery Office, 1983).

27. Donaghy, *British Government*, 92–93. There was confusion in the British Foreign Office as well. In a later interview, Rowlands recounted that, when a desk officer ran up to him, saying, "They've shot Shackleton!" he was initially confounded: "for a split second I actually thought that he meant they'd shot Eddie Shackleton. Of course it was in fact the RRS *Shackleton*, the research vessel named after his father, on which an Argentine destroyer had just opened fire. Indeed, it was a rather hectic afternoon." Rowlands quoted in Charlton, *Little Platoon*, 49.

28. "Argentines Fire Near British Ship," *Washington Post*, February 5, 1976; "British Ship Fired on by Argentine Warship," *Times* (London), February 5, 1976; Beck, "Cooperative Confrontation," 38–40; Donaghy, *British Government*, 92–93.

29. Freedman, *Official History*, 55; Jane Monahan, "British Envoy Returning from Argentina," *Times* (London), January 20, 1976. This belief persists in contemporary analyses of the incident. See, for example, Carolina Crisorio, "Malvinas en la política exterior Argentina," *Minius* 15 (2007): 67–83; and Jorge A. Fraga, "Petroleo en Malvinas: Cuestion de negocios o de soberania?," *Colección* 2 (1995): 115–24.

30. José Enrique Greño Velasco, "El 'informe Shackleton' sobre las islas Malvinas," *Revista de Política Internacional* 153 (1977): 38n25.

31. Donaghy, *British Government*, 89; James Nelson Goodsell, "Oil Issue: Who Rules Falklands?," *Christian Science Monitor*, December 30, 1975; Joanne Omang, "British and Argentines Step Up Island Dispute," *Washington Post*, January 22, 1976.

32. "Aha! Oil!," *Economist*, January 24, 1976.

33. Robert Lindley, "Argentine Minister Makes Plea for Economic Reason," *Financial Times*, February 11, 1976.

34. Charlton, *Little Platoon*, 50.

35. Dodds, *Pink Ice*, 151–52; Freedman, *Official History*, 22–23, 65.

36. Franks, *Falkland Islands Review*.

37. Virginia Gamba, *The Falklands/Malvinas War: A Model for North-South Crisis Prevention* (Winchester, MA: Allen and Unwin, 1987), 98–99.

38. Beck, *Falkland Islands*, 3; Donaghy, *British Government*, 92–94.

39. "Diplomacy to 'Cool' Argentine Incident," *Times*, February 6, 1976; Guillermo A. Makin, "Argentine Approaches to the Falklands/Malvinas: Was the Resort to Violence Foreseeable?," *International Affairs* 59, no. 3 (1983): 397.

40. "Diplomacy to 'Cool' Argentine Incident."

41. Quoted in Freedman, *Official History*, 39–41.

42. Quoted in Charlton, *Little Platoon*, 46.

43. "Islands in the Wind," *Economist*, December 24, 1977; Freedman, *Official History*, 40; O'Shaughnessy, "Decision."

44. Quoted in Freedman, *Official History*, 39. See also Freedman, 22–23; and Gustafson, *Sovereignty Dispute*, 83, 104.

45. Quoted in Charlton, *Little Platoon*, 58.

46. Charlton, 58–59.

47. Charlton, 55; Freedman, *Official History*, 79.

48. Charlton, *Little Platoon*, 68.

49. Donaghy, *British Government*, 65–66.

50. Hugh O'Shaughnessy, "Falkland Island May Be Rich in Offshore Oil and Fisheries," *Financial Times*, July 21, 1976.

51. "Islands in the Wind"; Hugh O'Shaughnessy, "Argentina and UK May Seek Falklands Oil," *Financial Times*, December 16, 1977.

52. Quoted in Michael Frenchman, "New Moves over the Falklands," *Times*, June 6, 1980.

53. Lawrence Freedman and Virginia Gamba-Stonehouse, *Signals of War: The Falklands Conflict of 1982* (Princeton, NJ: Princeton University Press, 1991), 12.

54. Freedman and Gamba-Stonehouse, 49–57.

55. Freedman, *Official History*, 187.

56. Charlton, *Little Platoon*, 114–16, 121; Freedman and Gamba-Stonehouse, *Signals*, 43–81, 142, 176, 241; Ned Lebow, "Miscalculation in the South Atlantic: The Origins of the Falklands War," *Journal of Strategic Studies* 6, no. 1 (1983): 5–35.

57. Caselli, Morelli, and Rohner, "Geography"; Safire, "In Defeat"; Westing, "Appendix 2," 208–9. Many Argentine scholars have also asserted that Britain's persistent commitment to holding the Falklands is driven by oil interests. For example, Alberto O. Casellas, *El territorio olvidado* (Buenos Aires: Centrol Naval, Instituto de Publicaciones Navales, 1974). British citizens, in turn, have accused Argentina of attempting to retake the islands in order to obtain more oil. For a discussion, see Peter Calvert, *The Falklands Crisis: The Rights and the Wrongs* (London: Frances Pinter, 1982), 46.

58. For an overview of the report, see Edward Shackleton, R. J. Storey, and R. Johnson, "Prospects of the Falkland Islands," *Geographical Journal* 143, no. 1 (1977): 1–13.

59. Michael Frenchman, "A Gleam of Hope at Last for Falkland Islanders," *Times*, December 13, 1977; Silenzi de Stagni, *Las Malvinas*, 82.

60. Michael Frenchman, "Fishing for a Way to End Island Dispute," *Times*, June 23, 1978; Michael Frenchman, "How Much Oil off the Falklands?," *Times*, December 17, 1979; Hugh O'Shaughnessy, "Cold Comfort from Latin America," *Financial Times*, February 14, 1979.

61. Freedman, *Official History*, 81.

62. Quoted in Freedman, 40–41.

63. Hugh O'Shaughnessy, "Argentina in Talks on Oil Development," *Financial Times*, March 15, 1977.

64. Based on data from the *BP Statistical Review of World Energy*, 67th ed. (London: BP, 2018). Argentina's natural gas reserves also tripled between 1976 and 1982. M. R. Yrigoyen, "The History of Hydrocarbon Exploration and Production in Argentina," *Journal of Petroleum Geology* 16, no. 4 (1993): 377.

65. Douglas Martin, "Oil Potential Is Minimized as Issue in Falklands Fight," *New York Times*, April 6, 1982.

66. Freedman, *Official History*, 40–41.

67. Charlton, *Little Platoon*, 101–2, 120–23; Freedman and Gamba-Stonehouse, *Signals*, 4, 9, 62, 68, 149; Lebow, "Miscalculation," 11–12, 20, 30; David A. Welch, "Remember the Falklands? Mixed Lessons of a Misunderstood War," *International Journal* 52, no. 3 (1997): 488.

68. For an articulation of the diversionary war hypothesis, see Jack S. Levy and Lily I. Vakili, "Diversionary Action by Authoritarian Regimes: Argentina in the Falklands/Malvinas Case," in *The Internationalization of Communal Strife*, ed. Manus I. Midlarsky (London: Routledge, 1992), 118–46. For a compelling rejection, see M. Taylor Fravel, "The Limits of Diversion: Rethinking Internal and External Conflict," *Security Studies* 19, no. 2 (2010): 307–41.

69. Welch, "Remember the Falklands," 486–87.

70. Klaus Dodds and Lara Manóvil, "A Common Space? The Falklands/Malvinas and the New Geopolitics of the South Atlantic," *Geopolitics* 6, no. 2 (2001): 109n45.

71. O'Shaughnessy, "Britain Denies."

72. Charlton, *Little Platoon*, 58–59.

73. Beck, *Falkland Islands*, 13; Freedman and Gamba-Stonehouse, *Signals*, 240–42.

7. OIL CAMPAIGNS

1. World War II, in particular, is commonly identified as an oil war. For example, see Brian C. Black, *Crude Reality: Petroleum in World History*, updated ed. (Lanham, MD: Rowman and Littlefield, 2014), 136–41; Andrew T. Price-Smith, *Oil, Illiberalism, and War: An Analysis of Energy and US Foreign Policy* (Boston: MIT Press, 2015), 76; and Paul Roberts, *The End of Oil: On the Edge of a Perilous New World* (Boston: Houghton Mifflin, 2005), 39.

2. "Memorandum on Japan's Oil Problem," *Memorandum of the Institute of Pacific Relations, American Council* 3 (December 21, 1934).

3. Irvine H. Anderson Jr., *The Standard-Vacuum Oil Company and United States East Asian Policy, 1933–1941* (Princeton, NJ: Princeton University Press, 1975), 226–27. Data are for 1939.

4. Christopher Thorne, *The Limits of Foreign Policy: The West, the League, and the Far Eastern Crisis of 1931–1933* (New York: G. F. Putnam and Sons, 1973), 63; Daniel Yergin, *The Prize: The Epic Quest for Oil, Money, and Power* (New York: Touchstone, 1992), 307.

5. Japan also invaded Burma, another oil producer, launching its main attack in January 1942 and seizing the state's oil fields at Yenangyaung in April 1942. However, Allied forces had significantly damaged the region's wells and refinery before the Japanese arrived. Robert Goralski and Russell W. Freeburg, *Oil and War: How the Deadly Struggle for Fuel in WWII Meant Victory or Defeat* (New York: William Morrow, 1987), 149, 152.

6. A similar argument is made by Daniel Moran, "The Battlefield and the Marketplace: Two Cautionary Tales," in *Energy Security and Global Politics: The Militarization of Resource Management*, ed. Daniel Moran and James A. Russell (New York: Routledge, 2009), 31.

7. Steven Ward, "Race, Status, and Japanese Revisionism in the Early 1930s," *Security Studies* 22, no. 4 (2013): 625–26.

8. Michael J. Barnhart, *Japan Prepares for Total War: The Search for Economic Security, 1919–1941* (Ithaca, NY: Cornell University Press, 1987), 28; Charles E. Neu, *The Troubled Encounter: The United States and Japan* (New York: John Wiley and Sons, 1975), 85. The United States recognized that Japan had "special interests" in China in the Lansing–Iishi exchange of notes (1917). However, this agreement was canceled during the Washington Naval Conference (1921–1922).

9. James B. Crowley, *Japan's Quest for Autonomy* (Princeton, NJ: Princeton University Press, 1966), 114–26; Shimada Toshihiko, "Designs on North China, 1933–1937," in *The China Quagmire: Japan's Expansion on the Asian Continent, 1933–1941*, ed. James William Morley (New York: Columbia University Press, 1983), 18–27.

10. For a detailed discussion of this time period, see Shimada, "Designs."

11. Crowley, *Japan's Quest for Autonomy*, 321, 333, 350; Dale C. Copeland, *Economic Interdependence and War* (Princeton, NJ: Princeton University Press, 2015), 148–49, 168.

12. Crowley, *Japan's Quest for Autonomy*, 321, 328–33, 344, 349–50; Hata Ikuhiko, "The Marco Polo Bridge Incident," in Morley, *China Quagmire*, 243–44, 250, 258–61, 266–68, 272–77. Copeland, *Economic Interdependence and War*, 169–73.

13. David Lu, introduction to Hata, "Marco Polo Bridge Incident," 237.

14. See Crowley, *Japan's Quest for Autonomy*, 225; Mark R. Peattie, "Nanshin: The 'Southward Advance,' 1931–1941, as a Prelude to the Japanese Occupation of Southeast Asia," in *The Japanese Wartime Empire, 1931–1945*, ed. Peter Duus, Ramon H. Myers, and Mark R. Peattie (Princeton, NJ: Princeton University Press, 1996), 195; Sandra Wilson,

"The Manchurian Crisis and Moderate Japanese Intellectuals: The Japan Council of the Institute of Pacific Relations," *Modern Asian Studies* 6, no. 3 (1992): 531; and the following documents in Joyce C. Lebra, *Japan's Greater East Asia Co-prosperity Sphere in World War II: Selected Readings and Documents* (Kuala Lumpur: Oxford University Press, 1975): "Konoye on the New Order in East Asia," 68–70; Matsuoka Yosuke, "Proclamation of the Greater East Asia Co-prosperity Sphere," 71–72; and Arita Hachiro, "The Greater East Asian Sphere of Common Prosperity," 73–77.

15. "East Asia Cooperative Body I," in Lebra, *Co-prosperity Sphere*, 11–12; Ian O. Lesser, *Resources and Strategy* (New York: St. Martin's, 1989), 86.

16. Fushun's annual crude oil output during the 1939–1941 period was reportedly 1–2.6 million barrels annually (143,000–371,000 tons). Helen Smyth, "China's Petroleum Industry," *Far Eastern Survey* 15, no. 12 (June 19, 1946): 187–90. Charles H. Behre Jr. and Kung-Ping Wang, "China's Mineral Wealth," *Foreign Affairs* 23, no. 1 (October 1944): 130–39, propose a higher figure but do not offer a source for it. Catherine Porter, "Mineral Deficiency versus Self-Sufficiency in Japan," *Far Eastern Survey* 5, no. 2 (January 15, 1936): 11, asserts that Fushun's output was not sufficient to meet Manchuria's demand, let alone Japan's. Anderson, *Standard-Vacuum*, 74, observes that, in 1934, Japan's total oil production from the home islands, Taiwan, Sakhalin Island, and Fushun met approximately 7 percent of domestic oil demand.

17. H. Foster Bain, "A Note on China's Petroleum Possibilities," *Far Eastern Survey* 15, no. 23 (November 20, 1946): 359; Behre and Wang, "China's Mineral Wealth"; Lesser, *Resources and Strategy*, 87, 89; Smyth, "China's Petroleum Industry," 188.

18. Walter J. Levy, *Oil Strategy and Politics, 1941–1981*, ed. Melvin A. Conant (Boulder, CO: Westview, 1982), 25, 28.

19. Goralski and Freeburg, *Oil and War*, 93.

20. Norman Angell, *Raw Materials, Population Pressure, and War* (Boston: World Peace Foundation, 1936), 24–25; Barnhart, *Japan Prepares*, 109–10; Alfred E. Eckes, *The United States and the Global Struggle for Minerals* (Austin: University of Texas Press, 1979), 73.

21. Herbert Feis, *The Road to Pearl Harbor: The Coming of War between the United States and Japan* (Princeton, NJ: Princeton University Press, 1950), 3.

22. Dorothy Borg, *The United States and the Far Eastern Crisis of 1933–1938: From the Manchurian Incident through the Initial Stage of the Undeclared Sino-Japanese War* (Cambridge, MA: Harvard University Press, 1964); Crowley, *Japan's Quest for Autonomy*, 165; Neu, *Troubled Encounter*, 138; Thorne, *Limits of Foreign Policy*, 15.

23. Robert Craigie, *Behind the Japanese Mask* (London: Hutchinson, 1945), 100. See also Peattie, "Nanshin," 238; and Tsunoda Jun, "The Decision for War," in *The Final Confrontation: Japan's Negotiations with the United States, 1941*, ed. James William Morley, trans. David A. Titus (New York: Columbia University Press, 1994), 243.

24. Anderson, *Standard-Vacuum*, 161; Goralski and Freeburg, *Oil and War*, 99.

25. Robert J. C. Butow, *Tojo and the Coming of the War* (Stanford, CA: Stanford University Press, 1961), 25, 45, 125, 223; Joseph C. Grew, *Turbulent Era: A Diplomatic Record of Forty Years* (London: Hammond, Hammond, 1953), 2:934, 1267; Nobutaka Ike, trans. and ed., *Japan's Decision for War: Records of the 1941 Policy Conferences* (Stanford, CA: Stanford University Press, 1967), 98; Neu, *Troubled Encounter*, 85, 103.

26. Akira Iriye, *Pacific Estrangement: Japanese and American Expansion, 1897–1911* (Cambridge, MA: Harvard University Press, 1972), 35–42, 50–58. Thompson and Dreyer date the beginning of Japan and the United States' rivalry to 1898. William R. Thompson and David R. Dreyer, *Handbook of International Rivalries, 1494–2010* (Washington, DC: CQ, 2012).

27. Thomas W. Burkman, *Japan and the League of Nations: Empire and World Order, 1914–1938* (Honolulu: University of Hawaii Press, 2008), 109; Butow, *Tojo*, 11, 19; Iriye,

Pacific Estrangement, 18–20, 42, 49–58, 133–36; Neu, *Troubled Encounter*, 49–52, 81–82, 101, 123; Usui Katsumi, "The Role of the Foreign Ministry," in *Pearl Harbor as History: Japanese-American Relations, 1931–1941*, ed. Dorothy Borg and Shumpei Okamoto, with the assistance of Dale K. A. Finlayson (New York: Columbia University Press, 1973), 139. Wilson, "Manchurian Crisis," 526, observes that moderate Japanese intellectuals considered the 1924 Immigration Act to be more of a detriment to positive US–Japan relations than Japan's occupation of Manchuria.

28. President Wilson resisted the clause because he believed it would increase the difficulty of treaty ratification. The British opposed it on behalf of Australian authorities, who were highly resistant to Japanese immigration. Burkman, *Japan*, 56, 80–85; Butow, *Tojo*, 18; Neu, *Troubled Encounter*, 100. On the racial discrimination issue more broadly, see Naoko Shimazu, *Japan, Race, and Equality: The Racial Equality Proposal of 1919* (London: Routledge, 2009); and Ward, "Race."

29. On the war scare, see Butow, *Tojo*, 11. On Hirohito's remarks, see David A. Titus, translator's introduction to Morley, *Final Confrontation*, xxxv.

30. For the quotation, see Crowley, *Japan's Quest for Autonomy*, 49. See also Crowley, 35, 43–44, 47–50, 52, 54–56; Neu, *Troubled Encounter*, 130; and Thorne, *Limits of Foreign Policy*, 37.

31. Stephen E. Pelz, *Race to Pearl Harbor: The Failure of the Second London Naval Conference and the Onset of World War II* (Cambridge, MA: Harvard University Press, 1974).

32. William G. Beasley, *Japanese Imperialism, 1894–1945* (Oxford: Oxford University Press, 1991), 211; Nazli Choucri, Robert C. North, and Susumu Yamakage, *The Challenge of Japan before World War II and After* (New York: Routledge, 1992), 132, 135; Akira Iriye, *After Imperialism: The Search for a New Order in the Far East, 1921–1931* (New York: Atheneum, 1969), 279.

33. Iriye, *After Imperialism*, 278–79; Yusuku Tsurumi, "Japan To-day and To-morrow," *International Affairs* 15, no. 6 (1936): 807–8.

34. Dale C. Copeland, "A Tragic Choice: Japanese Preventive Motivations and the Origins of the Pacific War," *International Interactions* 37, no. 1 (2011): 119; Peattie, "Nanshin," 195, 203; Jonathan G. Utley, *Going to War with Japan, 1937–1941* (New York: Fordham University Press, 2005), 91.

35. Yusuke Tsurumi, "Japan in the Modern World," *Foreign Affairs* 9, no. 2 (1931): 262. In the chapter's text, I present Japanese names using the conventional order of surname first. In the notes, I adhere to the format employed by each source. Some authors' names are therefore presented with surname first and some with surname second.

36. Grew, *Turbulent Era*, 930.

37. Grew, 934n930.

38. Butow, *Tojo*, 190; Pelz, *Race to Pearl Harbor*, 157, 190.

39. Iriye, *Pacific Estrangement*, 185, 195–96.

40. By December, the United States had extended $25 million in loans. "China Welcomes $25,000,000 Loan as Victory," *Chicago Tribune*, December 18, 1938; Grew, *Turbulent Era*, 1208.

41. Sumio Hatano and Sadao Asada, "Japan's Decision to 'Go South,'" in *Pearl Harbor and the Coming of the Pacific War: A Brief History with Documents and Essays*, ed. Akira Iriye (Boston: Bedford/St. Martin's, 1999), 126; Tsunoda Jun, "Confusion Arising from a Draft Understanding between Japan and the United States," in Morley, *Final Confrontation*, 4. Loans were offered in response to Japan joining the Tripartite Pact, but also in March 1940. "New 20 Million Dollar Loan Is Made to China," *Chicago Tribune*, March 8, 1940.

42. Jonathan Marshall, *To Have and Have Not: Southeast Asian Raw Materials and the Origins of the Pacific War* (Berkeley: University of California Press, 1995), 85.

43. Butow, *Tojo*, 125.

44. Barnhart, *Japan Prepares*, 153; Butow, *Tojo*, 153; Hata Ikuhiko, "The Army's Move into Northern Indochina," in *The Fateful Choice: Japan's Advance into Southeast Asia*, ed. James William Morley (New York: Columbia University Press, 1980), 155–58; Hatano and Asada, "Japan's Decision," 126, 134; Ike, *Japan's Decision for War*, xx, 212–14; Peattie, "Nanshin," 221–22; Robert A. Scalapino, introduction to Nagaoka Shinjiro, "Economic Demands on the Dutch East Indies," in Morley, *Fateful Choice*, 117. See also Japan's "General Principles to Cope with the Changing World Situation," accepted at the Liaison Conference, July 27, 1941.

45. Grew, *Turbulent Era*, 1225.

46. Sagan observes that a Japanese attack in summer 1940 was less likely to provoke a US response than one the next year. Scott D. Sagan, "The Origins of the Pacific War," *Journal of Interdisciplinary History* 18, no. 4 (1988): 898.

47. "General Principles of National Policy," Navy Headquarters, circa April 1936, in Lebra, *Co-prosperity Sphere*, 59–60.

48. "Summary Draft of a Policy for the South," Navy National Policy Research Committee, April 1939, in Lebra, *Co-prosperity Sphere*, 64–65. Leaders referred to "peaceful means" in the "Fundamentals of National Policy," issued at the Five Ministers' Conference, August 7, 1936, in Lebra, *Co-prosperity Sphere*, 36. The phrase was also repeated in a September 19, 1940, Imperial Conference. Ike, *Japan's Decision for War*, 8. Many observers also asserted that the navy's enthusiasm for a southern strategy was a "bureaucratic tactic" to counterbalance the army and secure larger budgets. Sadao Asada, "The Japanese Navy and the United States," in Borg and Okamoto, *Pearl Harbor as History*, 244; Peattie, "Nanshin," 213–17.

49. Tsunoda Jun, "The Navy's Role in the Southern Strategy," in Morley, *Fateful Choice*, 243–52.

50. Tsunoda, "Decision for War," 303.

51. David A. Deese, "Oil, War, and Grand Strategy," *Orbis* 25, no. 4 (1981): 539.

52. Butow, *Tojo*, 237; Feis, *Road to Pearl Harbor*, 105.

53. Tsunoda, "Navy's Role," 245–46, 251, 291–92.

54. Ike, *Japan's Decision for War*, 131, 153, 181; Tsunoda, "Decision for War," 258, 276, 292.

55. Quoted in Tsunoda Jun, "Leaning towards War," in Morley, *Final Confrontation*, 161–62.

56. Quoted in Tsunoda, "Decision for War," 287. The phrases appear in two notes, from September 29, 1941, and October 14, 1941.

57. Butow, *Tojo*, 237; Goralski and Freeburg, *Oil and War*, 94; Lesser, *Resources and Strategy*, 89–90; Tsunoda, "Decision for War," 277–78.

58. Seventy-five percent of planes on aircraft carriers were torpedo bombers and dive bombers, which had a range of about two hundred miles. Their attacks were able to disable enemy carriers.

59. Anderson, *Standard-Vacuum*, 156; H. Foster Bain, "Japan's Power of Resistance," *Foreign Affairs* 22, no. 3 (1944): 427; Goralski and Freeburg, *Oil and War*, 191–93.

60. Barnhart, *Japan Prepares*, 28–29; Levy, *Oil Strategy*, 29–30. Japan gave up the North Sakhalin concession in 1941 as part of its Neutrality Agreement with the USSR. Japan also acquired a concession in Dutch Borneo in 1930, but abandoned exploration because of insufficient funds. "Memorandum."

61. Barnhart, *Japan Prepares*, 102; "Memorandum."

62. Barnhart, *Japan Prepares*, 146; Goralski and Freeburg, *Oil and War*, 100; Anthony N. Stranges, "Synthetic Fuel Production in Prewar and World War II Japan: A Case Study in Technological Failure," *Annals of Science* 50 (1993): 229–65.

63. Goralski and Freeburg, *Oil and War*, 95; Nagaoka, "Economic Demands," 138.

64. The figures are from Anderson, *Standard-Vacuum*, 150–54, who quotes them in tons: 3.15 million tons and 1.85 million tons. I have used a standard conversion of seven barrels per ton. See also Barnhart, *Japan Prepares*, 165–66.

65. Barnhart, *Japan Prepares*, 207–8.

66. Anderson, *Standard-Vacuum*, 147–48; Butow, *Tojo*, 200; Nagaoka, "Economic Demands," 142–43.

67. Utley, *Going to War*, 143–45

68. US Department of State, *Papers Relating to the Foreign Relations of the United States (FRUS), Japan: 1931–1941* (Washington, DC: US Government Printing Office, 1943), 2:646, 676; Ike, *Japan's Decision for War*, 95–97, 270; Scalapino, introduction, 117.

69. Quoted in Ike, *Japan's Decision for War*, 246.

70. Quoted in Ike, 270.

71. Quoted in Ike, 236.

72. Butow, *Tojo*, 243, cites the figure of twelve thousand tons per day. See also Feis, *Road to Pearl Harbor*, 206–7, 261.

73. *FRUS, Japan: 1931–1941*, 2:714. At the Imperial Conference meeting that made the decision for war on December 1, 1941, Togo reiterated that, should Japan fail to stand up to the United States, "our very survival would be threatened." Ike, *Japan's Decision for War*, 270.

74. Grew, *Turbulent Era*, 1354, 1359–61.

75. *FRUS, Japan: 1931–1941*, 2:662. These suspicions intensified over the next six weeks. *FRUS, Japan: 1931–1941*, 2:712–14; Tsunoda, "Confusion," 102.

76. Sagan, "Origins of the Pacific War," 908–11.

77. *FRUS, Japan: 1931–1941*, 2:707–16; Tsunoda, "Decision for War," 266; Paul W. Schroeder, *The Axis Alliance and Japanese-American Relations, 1941* (Ithaca, NY: Cornell University Press, 1958), 156–67.

78. Schroeder, *Axis Alliance*, 89.

79. Quoted in Utley, *Going to War*, 173. There are two prominent, competing explanations for American intransigence. One emphasizes the information that the United States was receiving about Japan's war preparations. The United States had broken Japan's codes in September 1940, so authorities were aware that, as the Japanese pursued diplomatic negotiations, they were also preparing for war. US leaders interpreted this as evidence of Japanese perfidy and concluded that further negotiations were not worthwhile. Another explanation asserts that the Roosevelt administration was attempting to draw Japan into war and it was therefore the Americans who were negotiating in bad faith. For a variation of the latter argument, see John M. Schuessler, "The Deception Dividend: FDR's Undeclared War," *International Security* 34, no. 4 (2010): 133–65.

80. Quoted in Ike, *Japan's Decision for War*, 263.

81. Feis, *Road to Pearl Harbor*, 293. Feis is paraphrasing Tojo as the war decision was being made.

82. The phrase is from Levy, *Oil Strategy*, 34.

83. The first quotation is from Prince Konoe, speaking at an Imperial Conference on September 6, 1941; the second was expressed at a Liaison Conference on October 30, 1941. Ike, *Japan's Decision for War*, 138, 198.

84. Quoted in Ike, 152. Tojo later highlighted other perceived American threats, including Roosevelt's "arsenal of democracy" speech, mounting US expenditures on military expansion, the country's pursuit of a two-ocean fleet, expansion of the US Air Force, a new US base in Alaska, recommendations to evacuate US women and children from East Asia, and the proclamation of a state of national emergency. Butow, *Tojo*, 225.

85. The phrase is from Levy, *Oil Strategy*, 16.

86. Norman Rich, *Hitler's War Aims* (New York: Norton, 1974), 1:xxxi–xxxii.

87. Edward E. Ericson, *Feeding the German Eagle: Soviet Economic Aid to Nazi Germany, 1933–1941* (Westport, CT: Praeger, 1999), 24–25; Goralski and Freeburg, *Oil and War*, 17; Paul N. Hehn, *A Low Dishonest Decade: The Great Powers, Eastern Europe, and the Economic Origins of World War II, 1930–1941* (New York: Continuum, 2002), 14; Adolf Hitler, "Confidential Memo on Autarky, August 1936," in US Department of State, Division of Language Services, *Documents on German Foreign Policy: From the Archives of the German Foreign Ministry, 1918–1945*, ser. C, vol. 5 (Washington, DC: US Government Printing Office, 1966); Anand Toprani, "The Navy's Success Speaks for Itself: The German Navy's Independent Energy Security Strategy, 1932–1940," *Naval War College Review* 68, no. 3 (2015): 93; Yergin, *Prize*, 330.

88. Quoted in Arnold Krammer, "Fueling the Third Reich," *Technology and Culture* 19, no. 3 (1978): 399. See also Copeland, *Economic Interdependence and War*, 133–43; Hehn, *Low Dishonest Decade*, 14–15, 104; and Ian Kershaw, *Hitler: A Biography* (New York: W. W. Norton, 2008), 63–31. The "Hossbach Memorandum," which recorded a November 5, 1937, meeting between Hitler and his advisers, highlighted the risks posed by the foreign exchange issue and reliance on international trade. US Department of State, Division of Language Services, *Documents on German Foreign Policy: From the Archives of the German Foreign Ministry, 1918–1945*, ser. D, vol. 1 (Washington, DC: US Government Printing Office, 1949), 29–39.

89. Adolf Hitler, *Mein Kampf*, trans. James Vincent Murphy (London: Hutchinson, 1939); Woodruff D. Smith, *The Ideological Origins of Nazi Imperialism* (New York: Oxford University Press, 1989), 80, 92, 136–37, 217, 225–26.

90. Kershaw, *Hitler*, 446. See also Hitler, *Mein Kampf*, on the imperative for German expansion.

91. "Hossbach Memorandum"; Smith, *Ideological Origins*, 225–26.

92. "Petroleum Facilities of Austria" (prepared by the Enemy Oil Committee for the Division of Fuels and Lubricants, Office of the Quartermaster General, Washington DC, June 1944), 7–9. After the Anschluss, the German government transferred control over the industry to German companies. They implemented existing exploration plans, which resulted in significant new discoveries, including the Prinzendorf field. "Petroleum Facilities of Austria," 7, 29; Raymond G. Stokes, "The Oil Industry in Nazi Germany, 1936–1945," *Business History Review* 59, no. 2 (1985): 260–61. As a result, Austrian oil production increased to an estimated 1.24 million barrels in 1939, 2.48 million barrels in 1940, 2.72 million barrels in 1941, 3.8 million barrels in 1942, and 7.82 million barrels in 1943. "Petroleum Facilities of Austria," 8. Another report contains similar but not identical estimates. United States Strategic Bombing Survey (USSBS), *Oil Division: Final Report* (Washington, DC: US Government Printing Office, January 1947), 21.

93. Luděk Holub et al., *A Century of Petrol: The History of the Refining Industry in the Czech Lands* (Prague: Asco, 2005), 24. The figure they quote is thirty thousand tons.

94. Goralski and Freeburg, *Oil and War*, 30–31.

95. USSBS, *The German Oil Industry Ministerial Report, Team 78* (Washington, DC: US Government Printing Office, January 1947), 45. The original figure is sixty thousand to sixty-five thousand tons. Output was also delayed by French self-sabotage. Goralski and Freeburg, *Oil and War*, 34, 36–37.

96. USSBS, *German Oil Industry*, 70. The original figure is 906,000 tons; USSBS, *Oil Division*, 25. German forces captured few stocks in the other countries they conquered.

97. Goralski and Freeburg, *Oil and War*, 23–25.

98. The number is calculated from Levy, *Oil Strategy*, 15–16. He estimated that oil production, in all German-occupied territories, was approximately 75 million barrels in 1941 and that normal oil consumption for that year, without restrictions, would have been

200–210 million barrels. Levy, *Oil Strategy*, 10. Krammer, "Fueling the Third Reich," 409, reports that Axis Europe produced 84 million barrels (12 million tons) of oil in 1941. On the shortage issue, see also Ericson, *Feeding the German Eagle*, 116, 127; and Joel Hayward, "Hitler's Quest for Oil: The Impact of Economic Considerations for Military Strategy, 1941–42," *Journal of Strategic Studies* 18, no. 4 (1995): 101. Germany was aware of this problem, as it had started planning for a postwar New Order as soon as France fell. Anand Toprani, "Germany's Answer to Standard Oil: The Continental Oil Company and Nazi Grand Strategy, 1940–1942," *Journal of Strategic Studies* 37 (2014): 951.

99. On Soviet and Romanian responses, see Goralski and Freeburg, *Oil and War*, 23. Hitler had been aware of the blockade risk since at least 1937; he mentioned it in the meeting described in the "Hossbach Memorandum."

100. Howard Johnson, "The United States and the Establishment of the Anglo–American Caribbean Commission," *Journal of Caribbean History* 19, no. 1 (1984): 33; USSBS, *German Oil Industry*, 15; USSBS, *Oil Division*, 14. According to the last source, Germany's export dependence ranged from 70 to 89.5 percent between 1935 and 1938.

101. Krammer, "Fueling the Third Reich," 400; USSBS, *German Oil Industry*, 8, 20; USSBS, *Oil Division*, 14. The last provides the figures of 230,000 to 1 million tons. Stokes, "Oil Industry," 261–62, argues that tariffs aimed mostly to encourage synthetics; the German government had low expectations for German crude.

102. Goralski and Freeburg, *Oil and War*, 18–19, 21–23; Krammer, "Fueling the Third Reich," 395–98, 401; USSBS, *Oil Division*, 10; Yergin, *Prize*, 329–30.

103. Hitler, "Confidential Memo." See also Goralski and Freeburg, *Oil and War*, 25; and Yergin, *Prize*, 333, for alternative translations.

104. Krammer, "Fueling the Third Reich," 401–3.

105. USSBS, *German Oil Industry*, 21; USSBS, *Oil Division*, 14, 18, 21. The goal was 2.584 million tons, but actual output was 1.467 million tons.

106. USSBS, *Oil Division*, 21, reports that the synthetic fuels program provided 32 percent of total oil supplies in 1940 and 47 percent in 1944.

107. USSBS, *German Oil Industry*, 22; USSBS, *Oil Division*, 22.

108. Hayward, "Hitler's Quest for Oil," 117–18. The original figure is 8.7 million tons.

109. For a variety of perspectives on the agreement, see Hehn, *Low Dishonest Decade*, 235, 240–42; Kershaw, *Hitler*, 487; Maurice Pearton, *Oil and the Romanian State* (Oxford: Clarendon, 1971), 213–14, 219–20; Rich, *Hitler's War Aims*, 188; and Toprani, "Navy's Success," 94.

110. Ericson, *Feeding the German Eagle*, 89; Hehn, *Low Dishonest Decade*, 238, 284–85, 326–27, 335–36, 338–39, 355; Pearton, *Oil and the Romanian State*, 221, 244–45, 248; Rich, *Hitler's War Aims*, 188.

111. Pearton, *Oil and the Romanian State*, 221–23, 226–27, 251; Rich, *Hitler's War Aims*, 189.

112. Pearton, *Oil and the Romanian State*, 230, 232–37; Toprani, "Germany's Answer," 961.

113. Ericson, *Feeding the German Eagle*, 126; B. H. Liddell Hart, *History of the Second World War* (New York: G. P. Putnam's Sons, 1971), 143.

114. Hehn, *Low Dishonest Decade*, 354; Pearton, *Oil and the Romanian State*, 227; Rich, *Hitler's War Aims*, 189.

115. Kershaw, *Hitler*, 584; Pearton, *Oil and the Romanian State*, 223–24, 227–29, 233 (quotation on 228–29); Rich, *Hitler's War Aims*, 188–90, 206–7, 221; Norman Stone, *Hitler* (Boston: Little, Brown, 1980), 106–7.

116. The next year, Romania exported twenty-one million barrels (three million tons) of oil to Germany. USSBS, *German Oil Industry*, 46.

117. Toprani, "Germany's Answer," 952 (29 million tons). The surplus was 1.5 million tons (10.5 million barrels).

118. Ericson, *Feeding the German Eagle*, 16, 19–31; Richard Overy, *Russia's War: A History of the Soviet Effort, 1941–1945* (New York: Penguin Books, 1997), 37.

119. Ericson, *Feeding the German Eagle*, 28–36, 42, 44–50, 57–58, 61, 71, 105 (nine hundred thousand tons), 109–18. The Boundary and Friendship Treaty was signed on September 29 and backdated.

120. "Hossbach Memorandum."

121. Ericson, *Feeding the German Eagle*, 65–66, 89, 89, 102, 116, 126–27, 129, 134. The original figures are 34,000 tons, 155,000 tons, and 200,000 tons. See also Rich, *Hitler's War Aims*, 208. The Germans were partly to blame for the shortfalls, as some were retaliatory. However, by August 1940, Germany had fulfilled 55 percent of its deliveries, compared with the Soviets' 30 percent. Ericson, *Feeding the German Eagle*, 134.

122. Ericson, *Feeding the German Eagle*, 6–7, 128; Lesser, *Resources and Strategy*, 72; Levy, *Oil Strategy*, 17.

123. Hayward, "Hitler's Quest for Oil," 117–18. Original figures are from 8,701,000 tons to 5,577,000 tons.

124. Quoted in Goralski and Freeburg, *Oil and War*, 55.

125. Overy, *Russia's War*, 60; Liddell Hart, *History*, 143–44; Rich, *Hitler's War Aims*, 206.

126. The first quotation is from Overy, *Russia's War*, 61. The second appears in numerous sources, including Liddell Hart, *History*, 147–49; Rich, *Hitler's War Aims*, 207; and Yergin, *Prize*, 335.

127. H. R. Trevor-Roper, ed., *Hitler's War Directives, 1939–1945* (London: Sidgewick and Jackson, 1964), 48–52; Walter Warlimont, *Inside Hitler's Headquarters, 1939–45*, trans. R. H. Barry (New York: Frederick A. Praeger, 1964), 112–13, 135.

128. An August 1940 study by the Military Geography Department of the General Staff also highlighted "the fundamental importance of Baku as an operational objective." Horst Boog et al., *Germany and the Second World War*, vol. 6, *The Global War* (Oxford: Oxford University Press, 2001), xiii–iv. See also Hayward, "Hitler's Quest for Oil," 102; and Liddell Hart, *History*, 150.

129. Alex J. Kay, "German Economic Plans for the Occupied Soviet Union and Their Implementation, 1941–1944," in *Stalin and Europe: Imitation and Domination, 1928–1953*, ed. Timothy Snyder and Ray Brandon (Oxford: Oxford University Press, 2014).

130. The phrase is from Overy, *Russia's War*, 84.

131. Hitler, "Confidential Memo"; Yergin, *Prize*, 334. See also Ericson, *Feeding the German Eagle*, 127; Ian Kershaw, *Hitler, 1889–1936: Hubris* (New York: W. W. Norton, 1999), 243, 245–50, 288–89; Liddell Hart, *History*, 143; Rich, *Hitler's War Aims*, 212.

132. Ericson, *Feeding the German Eagle*, 12–13, 127–28; Hayward, "Hitler's Quest for Oil," 97; Hitler, *Mein Kampf*.

133. The quotation is from Halder's war diaries (July 31, 1940). Franz Halder, *The Halder War Diary, 1939–1942*, ed. Charles Burton Burdick and Hans-Adolf Jacobsen (Novato, CA: Presidio, 1988), 244. See also Ericson, *Feeding the German Eagle*, 127; Liddell Hart, *History*, 144; Overy, *Russia's War*, 566, 569, 588; Rich, *Hitler's War Aims*, 209; and Stone, *Hitler*, 108.

134. Trevor-Roper, *Hitler's War Directives*, 51. In the process, Germany would acquire Estonia's oil shale resources and the Soviet-controlled oil fields in Galicia. Goralski and Freeburg, *Oil and War*, 74; Krammer, "Fueling the Third Reich," 409; Toprani, "Navy's Success," 98. The former were heavily damaged by retreating Soviet forces, and the latter did not significantly increase German oil supplies, as the Soviets had been selling their share of Galicia's oil to Germany since September 1939.

135. Trevor-Roper, *Hitler's War Directives*, 85, 89.

136. Trevor-Roper, 93.

137. Quoted in Hayward, "Hitler's Quest for Oil," 107. See also Goralski and Freeburg, *Oil and War*, 78–79; Hayward, "Hitler's Quest for Oil," 95, 123–24; and Kershaw, *Hitler*, 641. Yergin, *Prize*, 334, asserts that 58 percent of Germany's oil came from Romania in 1940. Hayward, "Hitler's Quest for Oil," 99, states that 94 percent of the country's imports came from Romania.

138. Quoted in Trevor-Roper, *Hitler's War Directives*, 95. See also Goralski and Freeburg, *Oil and War*, 78; and Liddell Hart, *History*, 167.

139. Hayward, "Hitler's Quest for Oil," 104–5, 117; Liddell Hart, *History*, 168; USSBS, *Oil Division*, 28.

140. Hayward, "Hitler's Quest for Oil," 120–21; Kershaw, *Hitler*, 721; Liddell Hart, *History*, 245; Anand Toprani, "The First War for Oil: The Caucasus, German Strategy, and the Turning Point of the War on the Eastern Front, 1942," *Journal of Military History* 80 (2016): 824, 828, 835.

141. Quoted in Hayward, "Hitler's Quest for Oil," 118. See also Hayward, 102, 117.

142. Quoted in Joel Hayward, "Too Little, Too Late: An Analysis of Hitler's Failure in August 1942 to Damage Soviet Oil Production," *Journal of Military History* 64 (2000): 771. Also see Hayward, "Hitler's Quest for Oil," 95–96; and Toprani, "First War for Oil," 815, 839, 845–46, 854. The southern offensive was renamed Operation Brunswick at the end of June.

143. Quoted in Hayward, "Hitler's Quest for Oil," 125. See also Hayward, 95, 120, 123, 127; Kershaw, *Hitler*, 710–12; Overy, *Russia's War*, 157; Toprani, "First War for Oil," 827–30; and Trevor-Roper, *Hitler's War Directives*, 117–18, 131.

144. Trevor-Roper, *Hitler's War Directives*, 129–31.

145. Levy, *Oil Strategy*, 19–20; Hayward, "Hitler's Quest for Oil," 121–22; Toprani, "First War for Oil," 832–33, 841–45; Martin Van Creveld, *Supplying War: Logistics from Wallenstein to Patton* (Cambridge: Cambridge University Press, 1977), 150.

146. Goralski and Freeburg, *Oil and War*, 74–76; Directive 43, "Continuation of Operations from the Crimea," July 11, 1942, in Trevor-Roper, *Hitler's War Directives*.

147. Liddell Hart, *History*, 248–49.

148. Hayward, "Too Little, Too Late," 780.

149. Hayward, "Hitler's Quest for Oil," 135n151; Hayward, "Too Little, Too Late," 779–80; Toprani, "First War for Oil," 847–48.

150. Goralski and Freeburg, *Oil and War*, 182–84; Liddell Hart, *History*, 252; Toprani, "First War for Oil," 849.

8. OIL GAMBIT

1. For example, Daniel Deudney, "Environmental Security: A Critique," in *Contested Grounds: Security and Conflict in the New Environmental Politics*, ed. Daniel Deudney and Richard Anthony Matthew (Albany: State University of New York Press, 1999), 208; Daniel Moran and James A. Russell, "Introduction: The Militarization of Energy Security," in *Energy Security and Global Politics: The Militarization of Resource Management*, ed. Daniel Moran and James A. Russell (New York: Routledge, 2009), 17; and Richard H. Ullman, *Securing Europe* (Princeton, NJ: Princeton University Press, 1991), 25.

2. Erik Gartzke and Dominic Rohner, "To Conquer or Compel: War, Peace, and Economic Development" (Working Paper No. 511, Institute for Empirical Research in Economics, University of Zurich, September 2010), 7.

3. The quotation is from Janice Gross Stein, "Threat-Based Strategies of Conflict Management: Why Did They Fail in the Gulf?," in *The Political Psychology of the Gulf War: Leaders, Publics, and the Process of Conflict*, ed. Stanley A. Renshon (Pittsburg: University of Pittsburgh Press, 1993), 127–28, who challenges this view.

4. Charles Duelfer, *Comprehensive Report of the Special Advisor to the DCI on Iraq's WMD* (Washington, DC: Central Intelligence Agency, September 30, 2004), 1:42.

5. An example of this narrative appears in Efraim Karsh, "Rethinking the 1990–91 Gulf Conflict," *Diplomacy and Statecraft* 7, no. 3 (1996): 729–69.

6. F. Gregory Gause III, "Iraq's Decisions to Go to War, 1980 and 1990," *Middle East Journal* 56, no. 1 (2002): 47–48.

7. Saddam used the phrase "mother of all battles" to refer to the US-led response to his invasion of Kuwait.

8. Efraim Karsh and Inari Rautsi, "Why Saddam Hussein Invaded Kuwait," *Survival* 33, no. 1 (1991): 18.

9. Martin Viorst, "Report from Baghdad," *New Yorker*, September 24, 1990.

10. For 1989, calculated from "Iraq Economic Data (1989–2003)," US Central Intelligence Agency, April 23, 2007, https://www.cia.gov/library/reports/general-reports-1/iraq_wmd_2004/chap2_annxD.html.

11. Kiren Aziz Chaudhry, "On the Way to Market: Economic Liberalization and Iraq's Invasion of Kuwait," *Middle East Report*, no. 170 (1991): 14; Karsh and Rautsi, "Why Saddam Hussein Invaded," 19. Other authors estimate that Iraqi debt repayment cost $8 billion annually, while Iraqi prime minister Tariq ʿAziz reported $7 billion. Amatzia Baram, "The Iraqi Invasion of Kuwait: Decision Making in Baghdad," in *Iraq's Road to War*, ed. Amatzia Baram and Barry Rubin (Houndsmills, UK: Macmillan, 1994), 7; Viorst, "Report" (1990), 91.

12. Roland Dannreuther, *The Gulf Conflict: A Political and Strategic Analysis* (London: International Institute for Strategic Studies, 1991/1992), 10–11.

13. Tariq ʿAziz, interview in "The Gulf War: An In-Depth Examination of the 1990–1991 Persian Gulf Crisis," *Frontline*, originally aired on PBS, January 9, 1996, interview transcript, http://www.pbs.org/wgbh/pages/frontline/gulf/oral/iraqis.html. See also "Interview #15 (16 March)," in *Saddam Hussein Talks to the FBI: Twenty Interviews and Five Conversations with "High Value Detainee #1" in 2004*, ed. Joyce Battle, National Security Archive Electronic Briefing Book No. 279, http://www2.gwu.edu/~nsarchiv/NSAEBB/NSAEBB279/; Chaudhry, "On the Way to Market," 14, 20; Gause, "Iraq's Decisions," 58–59; and Karsh and Rautsi, "Why Saddam Hussein Invaded," 20.

14. Figures are for 1989. "Iraq Economic Data."

15. Karsh and Rautsi, "Why Saddam Hussein Invaded," 19; Janice Gross Stein, "Deterrence and Compellence in the Gulf, 1990–91: A Failed or Impossible Task?," *International Security* 17, no. 2 (1992): 158.

16. This figure was reported by Tariq ʿAziz. Viorst, "Report" (1990), 91.

17. Thomas C. Hayes, "Confrontation in the Gulf; the Oilfield Lying below the Iraq–Kuwait Dispute," *New York Times*, September 3, 1990. Together, Kuwait and the UAE were responsible for 75 percent of OPEC's overproduction. Lawrence Freedman and Efraim Karsh, *The Gulf Conflict, 1990–1991: Diplomacy and War in the New World Order* (Princeton, NJ: Princeton University Press, 1993), 41.

18. After the Iran–Iraq War the state's petroleum infrastructure also required significant investment before production could increase.

19. Chaudhry, "On the Way to Market," 15–18, 20; Fred H. Lawson, "Rethinking the Iraqi Invasion of Kuwait," *Review of International Affairs* 1, no. 1 (2007): 11–14.

20. Chaudhry, "On the Way to Market," 18, 20, 23; Freedman and Karsh, *Gulf Conflict*, 37–39; Lawson, "Rethinking," 13–14.

21. Baram, "Iraqi Invasion of Kuwait," 7; Karsh and Rautsi, "Why Saddam Hussein Invaded," 20; Lawson, "Rethinking," 15.

22. Chaudhry, "On the Way to Market," 18; Dannreuther, *Gulf Conflict*, 11–12; Majid Khadduri and Edmund Ghareeb, *War in the Gulf, 1990–91: The Iraq–Kuwait Conflict and Its Implications* (New York: Oxford University Press, 1997), 80; Lawson, "Rethinking," 10–11.

23. Gause makes a similar argument, observing that the economic crisis was perceived as "part of a more general play aimed at weakening and destabilizing the Baʿath regime, orchestrated from the outside." Gause, "Iraq's Decisions," 53.

24. ʿAziz, interview in "Gulf War"; "Interview Session #4 (13 February)" and "Interview Session #9 (24 February)," in Battle, *Saddam Hussein Talks*. This hypothesis also appears in Iraqi regime documents. For example, see "General Military Intelligence Directorate (GMID) Reports and Analysis about US Attacks against Iraq in 1993, GMID Role in Iraqi Battles, Um Al-Ma'arik, Qadissiyah Saddam," July 1991–September 2001, SH-GMID-D-000-513, Saddam Hussein Regime Collection, Conflict Records Research Center, National Defense University, Washington, DC; and "Report about the Iraqi–Kuwaiti Relations before and after Persian Gulf War," undated, SH-MISC-D-000-870, Saddam Hussein Regime Collection. These document, like others cited in this chapter, are from the Iraqi regime's archives and were acquired by US forces during the 2003 invasion of Iraq.

25. These assaults might be launched with Israeli cooperation. "Report by the Iraqi Military Intelligence Forces," May 22, 1990, SH-PDWN-D-000-546, Saddam Hussein Regime Collection. See also Hal Brands and David Palkki, "'Conspiring Bastards': Saddam Hussein's Strategic View of the United States," *Diplomatic History* 36, no. 3 (2012): 652.

26. Quoted in Gause, "Iraq's Decisions," 58.

27. ʿAziz, interview in "Gulf War"; Ofra Bengio, *Saddam's Word: Political Discourse in Iraq* (New York: Oxford University Press, 1998), 132–33, 136; "Lecture on the Iran–Iraq War," February 1987, SH-BATH-D-000-300, Saddam Hussein Regime Collection.

28. Hal Brands, "Making the Conspiracy Theorist a Prophet: Covert Action and the Contours of United States–Iraq Relations," *International History Review* 33, no. 3 (2011): 385, 392; "Study by Baʿth Party Division Command Regarding Economic Sanctions," 1990, SH-BATH-D-000-492, Saddam Hussein Regime Collection; "General Military Intelligence Directorate," SH-GMID-D-000-513.

29. Bengio, *Saddam's Word*, 69, 79–85, 146–47, 165–70; "Saddam Hussein Discusses Neighboring Countries and Their Regimes," undated (circa September 1980–November 1981), SH-SHTP-A-000-626, Saddam Hussein Regime Collection.

30. "Lecture on the Iran–Iraq War," SH-BATH-D-000-300; "Saddam Hussein Speech 1978 on Geopolitics, etc.," November 8, 1978, SH-PDWN-D-000-938, Saddam Hussein Regime Collection; Kevin M. Woods, *The Mother of All Battles: Saddam Hussein's Strategic Plan for the Persian Gulf War* (Annapolis, MD: Naval Institute Press, 2008), 32.

31. Quoted in Brands, "Conspiracy Theorist," 391. See also Brands, 386–93.

32. "General Military Intelligence Directorate," SH-GMID-D-000-513; "Iraqi Report on the Gulf War," undated (after September 1, 1991), SH-MISC-D-000-952, Saddam Hussein Regime Collection; "Cultural Methodology and Political Guidance; Study of Iran Iraq War," undated (circa 1990s), SH-MODX-D-000-487, Saddam Hussein Regime Collection; "President Saddam Hussein Attending a Meeting Regarding the Israeli Attack," undated (circa mid-June 1981), SH-SHTP-A-000-571, Saddam Hussein Regime Collection.

33. Hal Brands, "Inside the Iraqi State Records: Saddam Hussein, 'Irangate,' and the United States," *Journal of Strategic Studies* 34, no. 1 (2011): 95–118.

34. This quotation is from the official Iraqi transcript of the meeting, available in "Confrontation in the Gulf; Excerpts from Iraqi Document on Meeting with US Envoy," *New York Times*, September 23, 1990.

35. "Interview Session #4 (13 February)"; "Report by the Iraqi Military Intelligence Forces," SH-PDWN-D-000-546; "Saddam Hussein and Political Officials Discussing How to Deal with the Republican Guard and Other Issues Following the First Gulf War," undated, SH-SHTP-A-000-834, Saddam Hussein Regime Collection.

36. Saddam emphasized this point in his Arab Cooperation Council (ACC) speech and in a meeting with the US assistant secretary of state for Near East and South Asian affairs, John Kelly, earlier that month. For the transcript of the ACC speech, see "Iraq's Saddam Husayn," Amman Television Service in Arabic (1010 GMT, February 24, 1990), Foreign Broadcast Information Service (FBIS), Daily Report, Near East and South Asia, FBIS-NES-90-039, February 27, 1990, 1–5. On the Kelly meeting, see Don Obderdorfer, "Missed Signals in the Middle East," *Washington Post Magazine*, March 17, 1991. Saddam returned to the theme after the invasion. "Saddam Meeting with Yasir 'Arafat before the Gulf War," January 1991, SH-SHTP-A-000-611, Saddam Hussein Regime Collection.

37. Saddam also highlighted this issue in his ACC summit speech. However, this belief predated the ceasefire; in the mid-1980s, Saddam told his advisers that the United States wanted to establish bases in the Gulf. "Transcripts of Meetings between Saddam, Vice President of the RCC Izzat Ibrahim al-Tikriti, Minister of Defense Adnan Khairallah, and Army Chief of Staff Abd al-Jawad Zinun during the Iraq–Iran War," February 25, 1985–July 31, 1985, SH-SHTP-A-000-607, Saddam Hussein Regime Collection.

38. Robert Pear, "House Approves Sanctions against Iraq," *New York Times*, September 28, 1988. The two chambers ultimately failed to agree on a sanctions bill, despite overwhelming bipartisan support. William J. Eaton, "Sanctions over Iraq's Gas Use Die in Congress," *Los Angeles Times*, October 23, 1988.

39. Freedman and Karsh, *Gulf Conflict*, 25–28; Bruce W. Jentleson, *With Friends like These: Reagan, Bush, and Saddam, 1982–1990* (New York: W. W. Norton, 1994), 42, 123–31.

40. James A. Baker III, *The Politics of Diplomacy: Revolution, War and Peace, 1989–1992*, with Thomas M. DeFrank (New York: G. P. Putnam's Sons, 1995), 265. See also "Meeting between Saddam and the Soviet Delegation," October 1990, SH-PDWN-D-000-533, Saddam Hussein Regime Collection. Rumors of coup and assassination plots were rife in Iraq from 1988 to 1990, and there were reportedly at least four attempts on Saddam's life during this period. Freedman and Karsh, *Gulf Conflict*, 29–30.

41. Jentleson, *With Friends like These*, 132–37, 148.

42. John K. Cooley, "Pre-war Gulf Diplomacy," *Survival* 33, no. 2 (1991): 126; Jentleson, *With Friends like These*, 145–46; Khadduri and Ghareeb, *War in the Gulf*, 99; Obderdorfer, "Missed Signals"; Stein, "Deterrence and Compellence," 161–62.

43. Freedman and Karsh, *Gulf Conflict*, 33–35; Khadduri and Ghareeb, *War in the Gulf*, 99–100; Jentleson, *With Friends like These*, 113–15, 152, 158; Obderdorfer, "Missed Signals"; Stein, "Deterrence and Compellence," 163.

44. "Report by the Iraqi Military Intelligence Forces," SH-PDWN-D-000-546, presents an extended analysis of this issue.

45. Freedman and Karsh, *Gulf Conflict*, 32; Obderdorfer, "Missed Signals."

46. For the text of Saddam's April 1 speech, see "President Warns Israel, Criticizes U.S.," Baghdad Domestic Service in Arabic (1030 GMT, April 2, 1990), FBIS-NES-90-064, 32-36. In the speech, Saddam also railed against Bull's assassination and international reactions to the Bazoft execution.

47. Baker, *Politics of Diplomacy*, 270; Jentleson, *With Friends like These*, 155–59, 162.

48. Milton Viorst, "Report from Baghdad," *New Yorker*, June 24, 1991. The Iraqis regularly highlighted the cutoff in US agricultural exports as evidence of American hostility. "Study by Ba'th Party," SH-BATH-000-492; "Iraqi Report on the Gulf War," SH-MISC-D-000-952; "30 September 1990 Meeting between Saddam Hussein and Sheikh Sidi Ahmad Walad Baba, the Mauritanian Minister of Interior," September 30, 1990, SH-PDWN-D-000-467, Saddam Hussein Regime Collection; "Saddam and the Soviet Delegation," SH-PDWN-D-000-533; "Saddam Hussein and Political Officials," SH-SHTP-A-000-834.

49. The Kuwaiti portion of the field is known as Ratqa.

50. Freedman and Karsh, *Gulf Conflict*, 62.

51. Miriam Joyce, *Kuwait 1945–1996: An Anglo-American Perspective* (London: Frank Cass, 1998), 163–64.

52. Dannreuther, *Gulf Conflict*, 16; Mohamed Heikal, *Illusions of Triumph: An Arab View of the Gulf War* (London: Fontana, 1993), 228.

53. On the dispute's early history, see David H. Finnie, *Shifting Lines in the Sand: Kuwait's Elusive Frontier with Iraq* (Cambridge, MA: Harvard University Press, 1992), 12–39. For "nineteenth province," see F. Gregory Gause III, "Iraq and the Gulf War: Decision-Making in Baghdad," Columbia International Affairs Online Series (New York: Columbia University Press, 2001), 16, https://www.files.ethz.ch/isn/6844/doc_6846_290_en.pdf.

54. For examples of this material, see Richard Schofield, ed., *The Iraq–Kuwait Dispute*, vol. 6, *The International Status of Kuwait*, pt. 2, *1914–1994* (Southampton, UK: Archive Editions, 1994), 161–89, 221–38.

55. Schofield, 246–49.

56. Finnie, *Shifting Lines*, 110–11, 126–34; Schofield, *Iraq–Kuwait Dispute*, 502–7; H. V. F. Winstone and Zahra Freeth, *Kuwait: Prospect and Reality* (New York: Crane, Russak, 1972), 214–15.

57. Peter Mansfield, *Kuwait: Vanguard of the Gulf* (London: Hutchinson, 1990), 51. Qasim repeatedly insisted that he would not use force to press the territorial claim and pragmatically pointed out that if he had wanted to invade Kuwait, he would have done so immediately rather than publicly announcing the claim, thereby offering the United Kingdom an opportunity to fortify its position in the region. Mustafa M. Alani, *Operation Vantage: British Military Intervention in Kuwait, 1961* (Surrey: LAAM, 1990), 225–26.

58. Alani, *Operation Vantage*, 55–56, 63–68, 74–78, 128.

59. Schofield, *Iraq–Kuwait Dispute*, 246–49, 266–73, 293–98. As a British official wrote of as-Suwaidi's claim to the entirety of Kuwait, "It was clear from the rest of the text of the memorandum . . . that he did not really expect such arguments to be taken seriously." Richard Schofield, *Kuwait and Iraq: Historical Claims and Territorial Disputes* (London: Royal Institute of International Affairs, 1991), 79.

60. The port issue was increasingly pressing, as Iraq's one other Gulf outlet, the Shatt al-ʿArab, was approaching its shipping capacity. Khadduri and Ghareeb, *War in the Gulf*, 4649; Schofield, *Iraq–Kuwait Dispute*, 247–49; Schofield, *Kuwait and Iraq*, 78.

61. Tim Niblock, "Iraqi Policies towards the Arab States of the Gulf, 1958–1981," in *Iraq, the Contemporary State*, ed. Tim Niblock (London: Croom Helm, 1982), 129; Schofield, *Kuwait and Iraq*, 94–96.

62. Finnie, *Shifting Lines*, 153–54.

63. Peter Calvert, ed., *Border and Territorial Disputes of the World*, 4th ed. (London: John Harper, 2004), 472; Finnie, *Shifting Lines*, 153; Khadduri and Ghareeb, *War in the Gulf*, 73–75; Phebe Marr, *The Modern History of Iraq* (Boulder, CO: Westview, 1985), 221–22; Niblock, "Iraqi Policies," 143; Schofield, *Kuwait and Iraq*, 115–19.

64. Calvert, *Border and Territorial Disputes*, 472; Finnie, *Shifting Lines*, 158–62; Schofield, *Kuwait and Iraq*, 118–20, 123.

65. This included loans and transportation of Iraqi oil resources in Kuwaiti tankers. The latter action would precipitate the so-called Tanker War between Iran and the US Navy.

66. Baram, "Iraqi Invasion of Kuwait," 8.

67. Marion Farouk-Sluglett and Peter Sluglett, *Iraq since 1958: From Revolution to Dictatorship* (London: I.B. Tauris, 1990), 259.

68. Lawson, "Rethinking," 9.

69. Heikal, *Illusions of Triumph*, 207–9; Schofield, *Kuwait and Iraq*, 124–26.

70. Tariq ʿAziz told Iraqi journalist Saʿd al-Bazzaz that the invasion would give Iraq "a lung, open to the sea." Quoted in Musallam Ali Musallam, *The Iraqi Invasion of Kuwait:*

Saddam Hussein, His State, and International Power Politics (London: British Academic Press, 1996), 87.

71. Freedman and Karsh, *Gulf Conflict*, 62. See also Glenn Frankel, "Imperialist Legacy Lines in the Sand," *Washington Post*, August 31, 1990.

72. For a recent version of the miscalculation argument, see Charles A. Duelfer and Stephen Benedict Dyson, "Chronic Misperception and International Conflict: The U.S.-Iraq Experience," *International Security* 36, no. 1 (2011): 73–100.

73. Baker, *Politics of Diplomacy*, 360. See also "General Military Intelligence Directorate," SH-GMID-D-000-513.

74. ʿAziz, interview in "Gulf War." See also Viorst, "Report" (1991), 66–67.

75. Baram, "Iraqi Invasion of Kuwait," 9; "Meeting Between Saddam and Iraqi Officials Regarding the Arab Summit," undated (1989–1990), SH-SHTP-A-000-732, Saddam Hussein Regime Collection; Viorst, "Report" (1990), 91. Iraq's July 15 memorandum to the Arab League, mentioned later, also made the argument about other oil producers benefiting from the war. As the memorandum put it, "These funds found their way into the treasuries of the other oil-producing countries." Schofield, *Iraq–Kuwait Dispute*, 790.

76. Heikal, *Illusions of Triumph*, 209; Khadduri and Ghareeb, *War in the Gulf*, 85–87.

77. Dannreuther, *Gulf Conflict*, 15; Freedman and Karsh, *Gulf Conflict*, 45–46; Joseph Kostiner, "Kuwait: Confusing Friend and Foe," in Baram and Rubin, *Iraq's Road to War*, 107.

78. Text in "Saddam's 30 May Speech on OPEC Oil Guidelines," Baghdad Domestic Service in Arabic (0839 GMT, July 18, 1990), FBIS-NES-90-139, July 19, 1990, 21–22.

79. Freedman and Karsh, *Gulf Conflict*, 46–47.

80. Khadduri and Ghareeb, *War in the Gulf*, 106.

81. Baram, "Iraqi Invasion of Kuwait," 17; Heikal, *Illusions of Triumph*, 217–18; Kostiner, "Kuwait," 112; Viorst, "Report" (1991), 66.

82. Heikal, *Illusions of Triumph*, 218.

83. All quotations are from the memorandum, in Schofield, *Iraq–Kuwait Dispute*, 786–91.

84. Schofield, 792–94.

85. ʿAziz, interview in "Gulf War"; Viorst, "Report" (1990), 91–92.

86. Text in Schofield, *Iraq–Kuwait Dispute*, 788.

87. Text in "Tariq ʿAziz Reacts to Kuwaiti Government Memo," Baghdad INA in Arabic (0744 GMT, July 24, 1990), FBIS-NES-90-142, July 24, 1990, 23-24. Saddam made the same claim in his National Day speech of July 17, 1990.

88. Freedman and Karsh, *Gulf Conflict*, 51; Obderdorfer, "Missed Signals"; Stein, "Deterrence and Compellence," 151.

89. Obderdorfer, "Missed Signals."

90. "Saddam Hussein Addresses Visiting U.S. Senators," Baghdad Domestic Service in Arabic (1400 GMT, April 16 1990), FBIS-NES-90-074, April 17, 1990, 5-13; Obderdorfer, "Missed Signals."

91. Quoted in Woods, *Mother of All Battles*, 49.

92. The quotation is from the cable Ambassador Glaspie sent to Washington after the meeting, which was declassified through a Freedom of Information Act request submitted by William Safire. American Embassy Baghdad to Secretary of State, Washington DC, Telegram 04237, July 25, 1990, 1990BAGHDAD04237, http://www.washingtonpost.com/wp-srv/politics/documents/glaspie1-13.pdf?sid=ST2008040203634. Subsequently referenced as "Glaspie telegram."

93. "Glaspie telegram."

94. Heikal, *Illusions of Triumph*, 228.

95. Freedman and Karsh, *Gulf Conflict*, 56–57; F. Gregory Gause III, *The International Relations of the Persian Gulf* (Cambridge: Cambridge University Press, 2010), 98.

96. Stein, "Deterrence and Compellence," 156. Reports of the meeting are, unsurprisingly, mixed. For various accounts, see Freedman and Karsh, *Gulf Conflict*, 59–60; Gause, *International Relations*, 101; Heikal, *Illusions of Triumph*, 243–44; and Khadduri and Ghareeb, *War in the Gulf*, 114–17.

97. Freedman and Karsh, *Gulf Conflict*, 58; Obderdorfer, "Missed Signals."

98. Viorst, "Report" (1990), 91.

99. "Glaspie telegram"; "Iraq's Saddam Husayn," FBIS-NES-90-039.

100. Freedman and Karsh, *Gulf Conflict*, 49; Gause, "Iraq's Decisions," 54; Khadduri and Ghareeb, *War in the Gulf*, 122.

101. Heikal, *Illusions of Triumph*, 244.

102. Saddam made this offer in his meeting with Joseph Wilson, the deputy chief of mission at the US embassy in Baghdad, on August 6. Freedman and Karsh, *Gulf Conflict*, 91; Joseph C. Wilson, *The Politics of Truth: An Insider's Memoir* (New York: Carroll and Graf, 2004), 121. On September 1, Hammadi described how Kuwait's oil industry would be integrated into Iraq's. Karsh, "Rethinking," 750–51. In a statement on October 18, Saddam offered to sell oil at $21 per barrel. "A Statement about the Iraqi Sale of Oil at $21/barrel," in "Archives Manuscripts of Speeches and Declaration by Saddam and Military Spokesman Regarding the Invasion of Kuwait," August 1983–October 1990, SH-PDWN-D-000-929, Saddam Hussein Regime Collection.

103. "Interview Session #9 (24 February)"; "Saddam and His Senior Advisors Discussing Iraq's Historical Rights to Kuwait and the United States' Position," December 15, 1990, SH-SHTP-D-000-557, Saddam Hussein Regime Collection. In his interview for "Gulf War," ʿAziz also characterized the attack as "defensive."

104. Saddam saw little difference between himself and the Iraqi nation; in his mind, his regime falling was the equivalent of Iraq falling. Jerrold M. Post, "The Defining Moment of Saddam's Life: A Political Psychological Perspective on the Leadership and Decision Making of Saddam Hussein during the Gulf Crisis," in Renshon, *Political Psychology*, 52–53; Joseph Sassoon, *Saddam Hussein's Baʿth Party: Inside an Authoritarian Regime* (Cambridge: Cambridge University Press, 2012), 169, 173, 181, 191; Woods, *Mother of All Battles*, 33.

105. "Meetings between Saddam and the Yemeni President, and between Saddam and Dr. George Habash, Secretary General of the Popular Front of the Palestinian Liberation," August 1990–September 1990, SH-MISC-D-000-652, Saddam Hussein Regime Collection; "Saddam and the Soviet Delegation," SH-PDWN-D-000-533. These statements might be interpreted as Saddam's dishonest attempt to justify a greedy invasion. However, they are consistent with Iraqis' internal explanations for the event. In addition, the regime's records reveal that Saddam's public statements closely match the views he expressed in private. Stephen Benedict Dyson and Alexandra L. Raleigh, "Public and Private Beliefs of Political Leaders: Saddam Hussein in Front of a Crowd and behind Closed Doors," *Research and Politics* 1, no. 1 (2014): 1–7; Kevin M. Woods, David D. Palkki, and Mark E. Stout, eds., *The Saddam Tapes: The Inner Workings of a Tyrant's Regime, 1978–2001* (Cambridge: Cambridge University Press, 2011), 328.

106. ʿAziz, interview in "Gulf War."

107. ʿAziz interview.

108. "Saddam and the Yemeni President," SH-MISC-D-000-652; "Saddam Hussein and Political Officials," SH-SHTP-A-000-834; "Saddam Speaking about US, World, and Iraqi Politics," undated, SH-SHTP-A-000-836, Saddam Hussein Regime Collection.

109. Chaudhry, "On the Way to Market," 23, and Woods, Palkki, and Stout, *Saddam Tapes*, 166, use the term *desperation* to characterize the Iraqi invasion.

110. Max Boot, "A War for Oil? Not This Time," *New York Times*, February 13, 2003.

111. Neela Banerjee, "Arabs Have a Litmus Test for U.S. Handling of Iraqi Oil," *New York Times*, April 5, 2003.

112. John Esterbrook, "Rumsfeld: It Would Be a Short War," *CBS News*, November 15, 2002, https://www.cbsnews.com/news/rumsfeld-it-would-be-a-short-war/.

113. David Frum, "The Curse of Oil Dependence," *Jerusalem Post*, December 20, 2002.

114. Quoted in Philippe Le Billon and Fouad El Khatib, "From Free Oil to 'Freedom Oil': Terrorism, War and US Geopolitics in the Persian Gulf," *Geopolitics* 9, no. 1 (2004): 109.

115. For example, Michael Renner, "Post-Saddam Iraq: Linchpin of a New Oil Order," *Foreign Policy in Focus*, January 1, 2003.

116. Greg Muttitt, *Fuel on the Fire: Oil and Politics in Occupied Iraq* (New York: New Press, 2012), 38.

117. Michael R. Gordon and Bernard E. Trainor, *COBRA II: The Inside Story of the Invasion and Occupation of Iraq* (New York: Pantheon Books, 2006), 457–64.

118. EIPG briefings declassified by the Freedom of Information Division of the Executive Services Directorate in response to a Freedom of Information Act request by Greg Muttitt, https://www.esd.whs.mil/Portals/54/Documents/FOID/Reading%20Room/CPA_ORHA/10-L-0234_Iraqi_Energy_Policy_Records.pdf.

119. Bob Woodward, *Plan of Attack* (New York: Simon and Schuster, 2004). 322–23.

120. James Dao, "Navy Seals Easily Seize 2 Oil Sites," *New York Times*, March 22, 2003. See also Gordon and Trainor, *COBRA II*, 182–96.

121. Gordon and Trainor, *COBRA II*, 427.

122. On potential damage from shutdowns, see Edward P. Djerian and Frank G. Wisner, *Guiding Principles for U.S. Post-conflict Policy in Iraq* (New York: Council on Foreign Relations, 2003), 18.

123. John Cassidy, "Beneath the Sand," *New Yorker*, July 14, 2003; Peter Maass, *Crude World: The Violent Twilight of Oil* (New York: Alfred A. Knopf, 2009), 154–57.

124. Neela Banerjee, "Widespread Looting Leaves Iraq's Oil Industry in Ruins," *New York Times*, June 10, 2003.

125. Quoted in Muttitt, *Fuel on the Fire*, 67–68, 88.

126. Philippe Le Billon, "Corruption, Reconstruction and Oil Governance in Iraq," *Third World Quarterly* 26, no. 4–5 (2005): 702n51.

127. Muttitt, *Fuel on the Fire*; Nicole Weygandt, "Crude Choice: Diffusion of Developing World Oil Regimes" (working paper, Niehaus Center for Globalization and Governance, Princeton University, Princeton, NJ, 2017), 30–31.

128. Baker, *Politics of Diplomacy*, 436–37; George Bush and Brent Scowcroft, *A World Transformed* (New York: Alfred A. Knopf, 1998), 464, 489; Robert M. Gates, *Duty: Memoirs of a Secretary at War* (New York: Alfred A. Knopf, 2014), 26–27.

129. Quoted in James Mann, *Rise of the Vulcans: The History of Bush's War Cabinet* (New York: Viking, 2004), 190.

130. Cheney, in particular, repeatedly mentioned the obstacles to advancing to Baghdad in 1991. For examples, see Dick Cheney, *In My Time: A Personal and Political Memoir* (New York: Threshold Editions, 2011), 236–37; and Jeffrey Record, *Dark Victory: America's Second War against Iraq* (Annapolis, MD: Naval Institute Press, 2004). 3.

131. Muttitt, *Fuel on the Fire*, 31–34.

132. EIPG briefings.

133. Quoted in Woodward, *Plan of Attack*, 381. See also Woodward, 244, 258; and Gordon and Trainor, *COBRA II*, 166–67.

134. John S. Duffield, "Oil and the Iraq War: How the United States Could Have Expected to Benefit, and Might Still," *Middle East Review of International Affairs* 9, no. 2 (2005): 127.

135. Donald Rumsfeld, *Known and Unknown: A Memoir* (New York: Sentinel, 2011), 493.

136. EIPG briefings.

137. Djerian and Wisner, *Guiding Principles*, 17–18.

138. Djerian and Wisner, 2–4, 9, 17–18.

139. EIPG briefings.

140. Stephen Benedict Dyson, "What Really Happened in Planning for Postwar Iraq?," *Political Science Quarterly* 128, no. 3 (2013): 455–88; Rumsfeld, *Known and Unknown*, 450, 482–85, 488.

141. Muttitt, *Fuel on the Fire*, 89–90.

142. Cassidy, "Beneath the Sand." See also Steve Coll, *Private Empire: ExxonMobil and American Power* (New York: Penguin, 2012), 246–47; and Daniel Yergin, "A Crude View of the Crisis in Iraq," *Washington Post*, December 8, 2002.

143. Serge Schmemann, "Controlling Iraq's Oil Wouldn't Be Simple," *New York Times*, November 3, 2002.

144. For other rejections of this argument, see Nayna J. Jhaveri, "Petroimperialism: US Oil Interests and the Iraq War," *Antipode* 36, no. 1 (2004): 8; and Steve A. Yetiv, *Explaining Foreign Policy: US Decision-Making in the Gulf War*, 2nd ed. (Baltimore: Johns Hopkins University Press, 2011), 242–43.

145. Stuart W. Bowen Jr., *Hard Lessons: The Iraq Reconstruction Experience* (Arlington, VA: Office of the Special Inspector General for Iraq, 2009), 28–30.

146. Scholars that support this position include Toby Craig Jones, "America, Oil, and the Middle East," *Journal of American History* 99, no. 1 (2012): 210, 217; Muttitt, *Fuel on the Fire*, 25, 38; Andrew T. Price-Smith, *Oil, Illiberalism, and War: An Analysis of Energy and US Foreign Policy* (Boston: MIT Press, 2015), 91; and Doug Stokes, "Blood for Oil? Global Capital, Counter-insurgency and the Dual Logic of American Energy Security," *Review of International Studies* 33, no. 2 (2007): 251.

147. Michael T. Klare, *Blood and Oil: The Dangers and Consequences of America's Growing Dependency on Imported Petroleum* (New York: Henry Holt, 2004).

148. Price-Smith, *Oil, Illiberalism, and War*, 91.

149. For prominent versions of this argument, see Duffield, "Oil and the Iraq War"; Michael T. Klare, "For Oil and Empire? Rethinking the War with Iraq," *Current History* 102, no. 662 (2003): 129–35; Muttitt, *Fuel on the Fire*; and Price-Smith, *Oil, Illiberalism, and War*.

150. Robert E. Ebel, *The Geopolitics of Energy into the 21st Century* (Washington, DC: Center for Strategic and International Studies, 2000); Edward J. Morse and Amy Myers Jaffe, *Strategic Energy Policy: Challenges for the 21st Century*, Report of an Independent Task Force Cosponsored by the James A. Baker III Institute for Public Policy of Rice University and the Council on Foreign Relations (New York: Council on Foreign Relations, 2001); National Energy Policy Development Group, *National Energy Policy* (Washington, DC: US Government Printing Office, 2001).

151. On the impediments created by sanctions, see Fadhil Chalabi, "Iraq and the Future of World Oil," *Middle East Policy* 7, no. 4 (2000): 164–65.

152. "Vice President Speaks at VFW 103rd National Convention," August 26, 2003, George W. Bush White House, archived website, https://georgewbush-whitehouse.archives.gov/news/releases/2002/08/20020826.html.

153. Quoted in Price-Smith, *Oil, Illiberalism, and War*, 100.

154. Gause, *International Relations of the Persian Gulf*, 236. For the opposing perspective, see Jane K. Kramer and A. Trevor Thrall, "Introduction: Why Did the United States Invade Iraq?," in *Why Did the United States Invade Iraq?*, ed. Jane K. Kramer and A. Trevor Thrall (Milton Park, UK: Routledge, 2012), 1–24.

CONCLUSION

1. For a prominent example, see Kenneth S. Deffeyes, *Hubbert's Peak: The Impending World Oil Shortage* (Princeton, NJ: Princeton University Press, 2003).

2. The first predictions of peak oil demand appeared in 2013. Seth M. Kleinman et al., *Global Oil Demand Growth—The End Is Nigh* (New York: Citi Research, March 26, 2013).

3. The inverse is less likely, as the "shale revolution" has dramatically reduced US dependence on foreign energy resources.

4. Kejal Vyas, "Guyana Assures Exxon Venezuela Dispute Won't Slow Oil Exploration," *Wall Street Journal*, June 26, 2015.

5. On petro-myths and self-fulfilling prophecies, see Charles F. Doran, *Myth, Oil, and Politics: Introduction to the Political Economy of Petroleum* (New York: Free Press, 1977), 10.

6. Oil war skeptics make this claim, as described in chapter 1. However, this section's title is borrowed from Stephen Krasner, "Oil Is the Exception," *Foreign Policy* 14 (1974): 68–84, which argued that oil is exceptional in the sense that it is the one resource whose price could be manipulated by an international cartel.

7. Eugene Gholz and Daryl G. Press, "Protecting 'the Prize': Oil and the U.S. National Interest," *Security Studies* 19, no. 3 (2010): 453–58.

8. Over the last quarter century alone, the number of oil-producing states has doubled.

9. Aaron T. Wolf, "'Water Wars' and Water Reality: Conflict and Cooperation along International Waterways," in *Environmental Change, Adaptation, and Security*, ed. S. C. Lonergan (Dordrecht: Springer, 1999), 251–65.

10. Leonardo Maugeri, *The Age of Oil: The Mythology, History, and Future of the World's Most Controversial Resource* (Westport, CT: Praeger, 2006), xi.

11. Doran, *Myth, Oil, and Politics*; Sarah A. Emerson and Andrew C. Winner, "The Myth of Petroleum Independence and Foreign Policy Isolationism," *Washington Quarterly* 37, no. 1 (2014): 21–34; Eugene Gholz and Daryl G. Press, "Energy Alarmism: The Myths That Make Americans Worry about Oil" (Policy Analysis No. 589, CATO Institute, Washington, DC, 2007); Amy Myers Jaffe and Robert A. Manning, "The Myth of the Caspian 'Great Game': The Real Geopolitics of Oil," *Survival* 40, no. 4 (1998/1999): 112–29; Robin M. Mills, *The Myth of the Oil Crisis: Overcoming the Challenges of Depletion, Geopolitics, and Global Warming* (Westport, CT: Praeger, 2008); Robert Vitalis, *America's Kingdom: Mythmaking on the Saudi Frontier* (Stanford, CA: Stanford University Press, 2007); Steven A. Yetiv, *Myths of the Oil Boom: American National Security in a Global Energy Market* (Oxford: Oxford University Press, 2015).

12. Ida Tarbell, *The History of the Standard Oil Company* (New York: McClure, Philips, 1904).

13. William Cronon, "A Place for Stories: Nature, History, and Narrative," *Journal of American History* 78, no. 4 (1992): 1349.

14. On strategic narratives, see Alister Miskimmon, Ben O'Loughlin, and Laura Roselle, *Strategic Narratives: Communication Power and the New World Order* (New York: Routledge, 2013).

15. Molly Patterson and Kristen Renwick Monroe, "Narrative in Political Science," *Annual Review of Political Science* 1, no. 1 (1998): 320–21.

16. For example, Tor A. Benjaminsen, "Does Supply-Induced Scarcity Drive Violent Conflicts in the African Sahel? The Case of the Tuareg Rebellion in Northern Mali," *Journal of Peace Research* 45, no. 6 (2008): 819–36; Paul Richards, "Are 'Forest Wars' in Africa Resource Conflicts? The Case of Sierra Leone," in *Violent Environments*, ed. Nancy Lee Peluso and Michael Watts (Ithaca, NY: Cornell University Press, 2001); and Matthew D. Turner, "Political Ecology and the Moral Dimensions of 'Resource Conflicts': The Case of Farmer–Herder Conflicts in the Sahel," *Political Geography* 23, no. 7 (2004): 863–89.

17. Thad Dunning and Leslie Wirpsa, "Oil and the Political Economy of Conflict in Colombia and Beyond: A Linkages Approach," *Geopolitics* 9, no. 1 (2004): 81–108.

18. Séverine Autesserre, "Dangerous Tales: Dominant Narratives on the Congo and Their Unintended Consequences," *African Affairs* 111, no. 443 (2012): 202–22; Philippe Le Billon, "Digging into 'Resource War' Beliefs," *Human Geography* 5, no. 2 (2012): 26–40.

19. For an examination of this question applied to water, see David Katz, "Hydropolitical Hyperbole: Examining Incentives for Overemphasizing the Risks of Water Wars," *Global Environmental Politics* 11, no. 1 (2011): 12–35.

20. Michael Klare, "Resource Wars," *Harper's* 262, no. 1568 (1981): 20.

21. Ó Tuathail describes the process of generating storylines and its effects. Gearóid Ó Tuathail, "Theorizing Practical Geopolitical Reasoning: The Case of the United States' Response to the War in Bosnia," *Political Geography* 21, no. 5 (2002): 601–28. See also Shannon O'Lear, "Resource Concerns for Territorial Conflict," *GeoJournal* 64, no. 4 (2005): 298–300.

22. This is one of the concerns that motivate critiques of energy "securitization." Felix Ciută, "Conceptual Notes on Energy Security: Total or Banal Security?," *Security Dialogue* 41, no. 2 (2010): 123–44.

23. Jeff D. Colgan, "Fueling the Fire: Pathways from Oil to War," *International Security* 38, no. 2 (2013): 147–80.

24. Gholz and Press, "Protecting 'the Prize.'"

Index